Rights

Since 2013, an organization called the Nonhuman Rights Project has brought before the New York State courts an unusual request—asking for habeas corpus hearings to determine whether Kiko and Tommy, two captive chimpanzees, should be considered legal persons with the fundamental right to bodily liberty.

While the courts have agreed that chimpanzees share emotional, behavioural, and cognitive similarities with humans, they have denied that chimpanzees are persons on superficial and sometimes conflicting grounds. Consequently, Kiko and Tommy remain confined as legal "things" with no rights. The major moral and legal question remains unanswered: are chimpanzees mere "things", as the law currently sees them, or can they be "persons" possessing fundamental rights?

In *Chimpanzee Rights: The Philosophers' Brief*, a group of renowned philosophers considers these questions. Carefully and clearly, they examine the four lines of reasoning the courts have used to deny chimpanzees personhood: species, contract, community, and capacities. None of these, they argue, merits disqualifying chimpanzees from personhood. The authors conclude that when judges face the choice between seeing Kiko and Tommy as things and seeing them as persons—the only options under current law—they should conclude that Kiko and Tommy are persons who should therefore be protected from unlawful confinement "in keeping with the best philosophical standards of rational judgment and ethical standards of justice."

Chimpanzee Rights: The Philosophers' Brief—an extended version of the amicus brief submitted to the New York Court of Appeals in Kiko's and Tommy's cases—goes to the heart of fundamental issues concerning animal rights, personhood, and the question of human and nonhuman nature. It is essential reading for anyone interested in these issues.

Authors:

Kristin Andrews, York Research Chair in Animal Minds, Associate Professor of Philosophy, York University, Canada.

Gary Comstock, Professor of Philosophy, North Carolina State University, USA.

G.K.D. Crozier, Canada Research Chair in Environment, Culture and Values, Professor of Philosophy, Laurentian University, Canada.

Sue Donaldson, Research Associate, Department of Philosophy, Queen's University, Canada.

Andrew Fenton, Assistant Professor of Philosophy, Dalhousie University, Canada.

Tyler M. John, Ph.D. Student in Philosophy, Rutgers University, USA.

L. Syd M Johnson, Associate Professor of Philosophy and Bioethics, Michigan Technological University, USA.

Robert C. Jones, Associate Professor of Philosophy, California State University, Chico, USA.

Will Kymlicka, Canada Research Chair in Political Philosophy, Queen's University, Canada.

Letitia Meynell, Associate Professor of Philosophy, Dalhousie University, Canada.

Nathan Nobis, Associate Professor of Philosophy, Morehouse College, USA.

David Peña-Guzmán, Assistant Professor of Humanities and Liberal Studies, California State University, San Francisco, USA.

Jeffrey Sebo, Clinical Assistant Professor of Environmental Studies, Affiliated Professor of Bioethics, Medical Ethics, and Philosophy, and Director of the Animal Studies M.A. Program, New York University, USA.

Foreword: Lori Gruen is William Griffin Professor of Philosophy at Wesleyan University, USA, coordinator of the Wesleyan Animal Studies program, and Professor of Feminist, Gender and Sexuality Studies and Science in Society.

Afterword: Steven M. Wise is an American legal scholar, a former president of the Animal Legal Defense Fund, and founder and president of the Nonhuman Rights Project.

KRISTIN ANDREWS
GARY COMSTOCK
G.K.D. CROZIER
SUE DONALDSON
ANDREW FENTON
TYLER M. JOHN
L. SYD M JOHNSON
ROBERT C. JONES
WILL KYMLICKA
LETITIA MEYNELL
NATHAN NOBIS
DAVID PEÑA-GUZMÁN
JEFF SEBO

Chimpanzee Rights

THE PHILOSOPHERS' BRIEF

Routledge
Taylor & Francis Group
LONDON AND NEW YORK

First published 2019
by Routledge
2 Park Square, Milton Park, Abingdon, Oxon OX14 4RN

and by Routledge
711 Third Avenue, New York, NY 10017

Routledge is an imprint of the Taylor & Francis Group, an informa business

© 2019 Kristin Andrews, Gary Comstock, G.K.D. Crozier, Sue Donaldson, Andrew Fenton, Tyler M. John, L. Syd M Johnson, Robert C. Jones, Will Kymlicka, Letitia Meynell, Nathan Nobis, David Peña-Guzmán, and Jeff Sebo

Foreword © 2019 Lori Gruen
Epilogue © 2019 Steven M. Wise

The right of Kristin Andrews, Gary Comstock, G.K.D. Crozier, Sue Donaldson, Andrew Fenton, Tyler M. John, L. Syd M Johnson, Robert C. Jones, Will Kymlicka, Letitia Meynell, Nathan Nobis, David Peña-Guzmán, and Jeff Sebo to be identified as authors of this work has been asserted by them in accordance with sections 77 and 78 of the Copyright, Designs and Patents Act 1988.

All rights reserved. No part of this book may be reprinted or reproduced or utilised in any form or by any electronic, mechanical, or other means, now known or hereafter invented, including photocopying and recording, or in any information storage or retrieval system, without permission in writing from the publishers.

Trademark notice: Product or corporate names may be trademarks or registered trademarks, and are used only for identification and explanation without intent to infringe.

British Library Cataloguing-in-Publication Data
A catalogue record for this book is available from the British Library

Library of Congress Cataloging-in-Publication Data
A catalog record has been requested for this book

ISBN: 978-1-138-61863-3 (hbk)
ISBN: 978-1-138-61866-4 (pbk)
ISBN: 978-0-429-46107-1 (ebk)

Typeset in Joanna
by codeMantra

For Kiko and Tommy

Contents

Acknowledgments		xi
Foreword *Lori Gruen*		xiii
Introduction: Chimpanzees, rights, and conceptions of personhood	One	1
The species membership conception	Two	13
The social contract conception	Three	41
The community membership conception	Four	61
The capacities conception	Five	77
Conclusions	Six	101
Afterword *Steven M. Wise*		111
Author index		117
Subject index		119

Acknowledgments

The authors thank the many people who have helped us throughout the development of this book. James Rocha, Bernard Rollin, Adam Shriver, and Rebecca Walker were fellow travelers with us on the amicus brief but were unable to follow us to the book. Research assistants Andrew Lopez and Caroline Vardigans provided invaluable support and assistance at crucial moments. We have also benefited from discussion with audiences at the Stanford Law School and Dalhousie Philosophy Department Colloquium where the amicus brief was presented, and from the advice of wise colleagues, including Charlotte Blattner, Matthew Herder, Syl Ko, Tim Krahn, and Gordon McOuat. Lauren Choplin, Kevin Schneider, and Steven Wise patiently helped us navigate the legal landscape as we worked on the brief, related media articles, and the book. They continue to fight for freedom for Kiko, Tommy, and many other nonhuman animals.

Foreword
Lori Gruen

Anybody who has spent time with or seen movies of young chimpanzees knows just how endearing they can be. They are funny, playful, mischievous, and entertaining. That they are so alluring when young, is probably why Tommy and Kiko ended up in such awful conditions. As adorable young chimpanzees grow into the strong wild animals that they become as adults, many bought as pets end up in small cages, in garages or basements. The people who own these chimpanzees usually cannot care for them and the chimpanzees are left to suffer terribly, alone, often for decades.

I've met a number of chimpanzees who have once been pets or used for entertainment who were given up or rescued from their sad situations and brought to sanctuaries. Fortunately, the numbers of chimpanzees held in these debilitating forms of captivity are dwindling. In addition to Tommy and Kiko, there are roughly 35 chimpanzees held as pets and for entertainment in the United States, but there are another 150 or so being held in unaccredited facilities, like roadside zoos and wildlife refuges.[1] These facilities house chimpanzees in various conditions, but almost none of these places provide chimpanzees what they need. In contrast, there are six chimpanzee sanctuaries in the United States that are accredited by the Global Federation of Animal Sanctuaries and two others that are founding members of the North American Primate

Sanctuary Alliance. All eight sanctuaries do everything they can to care for these highly intelligent, emotionally sophisticated individuals. Chimpanzees who live in "true" sanctuaries are provided with the best food, veterinary care, and social and intellectual enrichment as certified by doctors, veterinarians, and chimpanzee experts in a transparent process. These sanctuaries are committed to providing excellent species-appropriate indoor and outdoor housing, allowing chimpanzees to live with other chimpanzees and make choices about what to do and whom to spend time with.[2]

Many of the chimpanzees who have moved to sanctuary have recovered from the trauma of their early physical and social deprivation, but some still bear the scars they accumulated in their early encounters with humans. Fortunately, most of those rescued, like Bubbles, who once was Michael Jackson's pet, or Marco, one of the oldest male chimpanzees in the country, who once lived in a cage that didn't allow him to stand up, adjust to life with other chimpanzees, in this case at the Center for Great Apes in Wachula, Florida, where they continue to thrive.[3] Henry, a former pet now living at Chimp Haven in Shreveport, Louisiana, has grown into the leader of his chimpanzee family. Rescued by the Houston SPCA after he was found severely undernourished in a rusty, feces-encrusted cage in a garage, he was nursed back to health, brought to the sanctuary, introduced to other chimpanzees, and is now a dominant, flourishing individual.[4]

Henry and I share a deep fondness for an older chimpanzee, Sarah. Sarah helped Henry understand chimpanzee behavior and taught him a few tricks too. I'll never forget the time I brought Sarah some of her favorite treats. Henry and Sarah were sitting on a perch in their play yard when Sarah motioned for me to quickly come to the mesh that safely separated us and give her the treats. Usually when chimpanzees see food they like, they "food grunt." These vocalizations can attract the attention of other chimpanzees, and, on this occasion, just as Henry was about to make noise, Sarah put her hand over his mouth so that the two of them could enjoy the treats I brought without having to share with others.

Sarah is basking in her retirement after a long career in cognition research aimed to explore whether chimpanzees have beliefs, desires, and intentions, can attribute those to others, can use language, solve problems, and engage in other highly sophisticated mental tasks, some of which are discussed in the pages that follow. She was one of the chimpanzee pioneers in this research,[5] and after more than four decades, she is now able to do what she wants—whether it is relaxing in her hammock, spending time with Henry and others, or demanding treats from her caregivers and friends like me.

When I first met Sarah, she and a group of eight other chimpanzees lived at the Chimp Center at Ohio State University, where Dr. Sally Boysen was engaged in cognition research with the group. Four of the younger chimps at the Center were "enculturated" which means that they were raised from infancy by Dr. Boysen and other care givers and were always treated as if they were human infants. Their early development was not unlike that of human children, and they were quite exceptional in their emotional and cognitive abilities. While this form of rearing is in no way ideal, as chimpanzees should grow up to learn to be chimpanzees, this particular enculturation experiment did allow the chimpanzees to live with other chimpanzees as well as to interact closely with humans. One of the dangers of this sort of interspecies relationship, however, is that the chimpanzees can become quite dependent on their human caregivers to help them navigate the complexities of chimpanzee social relations. This is what happened with this group and it led to tragedy.

Ohio State University decided to close the Chimp Center and ship the chimpanzees to a non-accredited refuge that was ill-equipped to care for this unique group. The chimps were separated from their human caregivers and from each other, and two of the adult males, Kermit and Bobby, died. Word got out that Sarah's health was deteriorating and the situation looked dire. Because chimpanzees are considered legal "property" and Dr. Boysen did not own them, there were no legal grounds to save them. For nine harrowing months, we pursued every option and, finally, through a coordinated effort of lawyers, scholars, activists, and chimpanzee experts, we were able to move the six remaining chimpanzees, including Sarah, to Chimp Haven, the National Chimpanzee Sanctuary. It is currently the largest chimpanzee sanctuary in the world and will be home to all of the federally owned chimpanzees as they retire from laboratories over the next few years.

I felt fortunate to be able to help save this group and help them move to Chimp Haven, as I had grown quite fond of these chimpanzees. In addition to Sarah, I developed a particularly close relationship with Emma, the youngest chimpanzee in the group, who was only five years old when this tragic saga began.

As I mentioned, young chimpanzees are adorably endearing, but that wasn't the only explanation for the bond I developed with Emma. Actually, it was Emma's distinctive personality and care for me that drew me in. One afternoon at the Chimp Center before it closed, Dr. Boysen, a few of Emma's caregivers, her young chimp companion, Harper, and my students and I went for a walk; Emma ran to me and immediately jumped into my arms. She would occasionally get down to play with Harper or look into my pockets, but she never left my side and either held my hand as we walked or climbed back into my arms. We interacted a lot after that walk, and it is an understatement to say that we formed a strong connection, one that clearly compelled me to do all I could to rescue her and the others from the dangerous situation they ended up in.

My relationship with Emma over the years has helped me think differently about my relationships with other chimpanzees, some of whom I know, and some, like Tommy and Kiko, whom I don't. This relationship has also changed the work I do as a philosopher. In particular, it pushed me to reflect on the skills we develop in the relationships we have with those who are quite different from us, and how those skills can help us expand our moral perception in order to see and value others who are distant in time or space or different in other ways, perhaps based on class, or race, or gender expression, or religion, or ability, or species. My relationship with Emma helped shape my view about entangled empathy as an ethical process for navigating difference.[6]

If someone asked me whether Emma is a person or a thing, I would be flabbergasted. She is an individual with whom I have developed a meaningful, caring relationship. Though she and I live very different lives and have distinct, independent points of view, we can come to share perspectives in our interactions and in that process we can change each other. I have known Emma for over a decade now, and she still does things that surprise me. Once, as I was leaving the sanctuary, she gathered up a bouquet of plants from inside her play yard and passed it to me through the mesh. People who have worked daily with chimpanzees for their entire careers were as amazed as I was touched. Is that what happens in relationships with persons or with things? This sort of question reflects a limited capacity for moral perception, one that fails to see individuals and their relationships for what they are. Emma, cannot be reduced to a set of capacities or to the remarkable activities that she engages in—none of us can.

Perhaps, following philosopher John Locke, who defines a person as a thinking thing (see page 86), we don't have to make a choice to answer the question. We are persons and things—thinking, feeling things in relationships with other thinking, feeling things. Unfortunately, under the law, the option of rejecting the division between person and thing is not yet available and this vexing question is one that must be answered. My colleagues in this volume, through careful argumentation, provide the answer that Emma, Sarah, Henry, Kiko, Tommy, and all chimpanzees are legal persons. They discuss the variety of ways personhood is understood and uncover a set of problems that led one legal scholar to note the dumbfounding way that "*personhood* becomes the term through which humans turn into slaves, ghosts turn into persons, or the biologically alive turn into the legally dead."[7] I am struck in reading the arguments here by the fact that while this odd division between person and thing is meant to simplify the law, it seems to make things more complicated for jurists. The arguments of the philosophers' amicus curiae brief and further elaborated in the chapters that follow clearly show that the judges in the case of Tommy and Kiko relied on "an inconsistent and, at times, incoherent understanding of what it means to be a person." (p. 116 of this volume)

At the heart of the philosophical, ethical, and legal force of the notion of personhood is a particular status. Status categories are always shaped by prevailing social norms and attitudes. These social norms have not only excluded chimpanzees and other animals from legal recognition, they end up condoning the violence that animals are forced to endure. In addition to clarifying the concept of personhood, *The Philosophers' Brief* helps to reshape these norms and attitudes. And already one judge has noticed: "While it may be arguable that a chimpanzee is not a "person," there is no doubt that it is not merely a thing (sic)."[8] The legal case for elevating the status of chimpanzees has not yet been accepted by any court in the United States, but the ethical case is clear. Emma, Sarah, Kiko, Tommy, and other chimpanzees deserve our respect, and we can show them this respect by supporting those who provide them with the very best care in sanctuaries.

NOTES

1 For up-to-date numbers please see the ChimpCare website: www.chimpcare.org/map.
2 See www.primatesanctuaries.org/about/member-sanctuaries/ and www.sanctuaryfederation.org/for-sanctuaries-2/.
3 See www.centerforgreatapes.org/meet-apes/chimpanzees/.
4 See www.chimphaven.org/chimps/henry/.
5 I have written about Sarah and her role in cognition research in Lori Gruen (2011). *Ethics and Animals: An Introduction*. Cambridge University Press.
6 See *Entangled Empathy: An Alternative Ethic for our Relationships with Animals* (2015). Lantern Press.
7 Colin Dayan (2011). *The Law Is a White Dog: How Legal Rituals Make and Unmake Persons*. Princeton University Press, p. 20.
8 Appeals Court ruling, Justice Fahey in "The Matter of Nonhuman Rights Project, Inc. v. Lavery."

Introduction: Chimpanzees, rights, and conceptions of personhood

One

In December 2013, the Nonhuman Rights Project (NhRP) filed a petition for a common law writ of habeas corpus in the New York State Supreme Court on behalf of Tommy, a chimpanzee living alone in a cage in a shed in rural New York (Barlow 2017). Under animal welfare laws, Tommy's owners, the Laverys, were doing nothing illegal by keeping him in those conditions. Nonetheless, the NhRP argued that given the cognitive, social, and emotional capacities of chimpanzees, Tommy's confinement constituted a profound wrong that demanded remedy by the courts. Soon thereafter, the NhRP filed habeas corpus petitions on behalf of Kiko, another chimpanzee housed alone in Niagara Falls, and Hercules and Leo, two chimpanzees held in research facilities at Stony Brook University. Thus began the legal struggle to move these chimpanzees from captivity to a sanctuary, an effort that has led the NhRP to argue in multiple courts before multiple judges.[1] The central point of contention has been whether Tommy, Kiko, Hercules, and Leo have legal rights. To date, no judge has been willing to issue a writ of habeas corpus on their behalf. Such a ruling would mean that these chimpanzees have rights that confinement might violate. Instead, the judges have argued that chimpanzees cannot be bearers of legal rights because they are not and cannot be *persons*.

Under current U.S. law, every entity is either a person or a thing. Those exhaust the legal choices. Since habeas corpus applies only to persons, as legal "things," Tommy, Kiko, Hercules, and Leo—like all captive chimpanzees—can be owned, moved, or sold at their owners' discretion (barring violations of animal welfare laws and subject to the Endangered Species Act). At the time of this writing, Kiko and Tommy remain captive in New York State. Hercules and Leo were moved from Stony Brook University (which had them on loan) to the New Iberia Research Center in Louisiana. In March 2018, they were transferred to Project Chimps, a chimpanzee sanctuary in Georgia (Grimm 2018). So, the current cases before the courts concern only Kiko and Tommy.

The process leading up to these petitions and the first two years of the subsequent legal battle are the subject of the 2016 documentary, *Unlocking the Cage*. The film has been screened and broadcast internationally, sometimes with the NhRP's founder, president, and chief litigator, Steven Wise, present. After one screening, Steven and two of the authors of this book, Letitia Meynell and Andrew Fenton, met to discuss some of the conceptual and logical problems with the courts' rulings. Within a few months, a team of seventeen philosophers assembled to collaborate on an amicus curiae brief in support of the NhRP's appeal of Kiko's and Tommy's cases. This book is an expanded version of that brief, which was submitted to the New York Court of Appeals in February 2018.

PHILOSOPHERS AND THE LAW

If you have ever had a conversation with a philosopher, you might be surprised that it is possible to get a group of philosophers to agree about anything. The seventeen of us are trained in different philosophical traditions and different styles of argument. We work in different subdisciplines of philosophy, including bioethics, philosophy of mind, political philosophy, and philosophy of science. We bring diverse views on the nature of personhood to this project. Some of us think that 'person' may not be a philosophically functional concept, and suspect it might be better if philosophers stopped using it. Despite our differences, what unites us all is the conviction that the current conditions in which Kiko and Tommy are held captive constitute an outrageous injustice. They are not mere things and to treat them as if they are is morally obscene.

Another thing that unites us is the belief that the discipline of philosophy has a vital role to play in debates about defining personhood and determining who should be counted as persons. 'Person' and 'personhood' have meant many different things at different times to different people. A survey of the complete global history of the concept would fill several large books. In recent European and American legal and philosophical thought, 'person' is primarily a moral concept designating an individual's moral standing (Chan and Harris 2011). In legal settings, personhood matters because it is taken to identify beings with sufficiently high moral standing to deserve protection under law.

As we read the various rulings in Kiko's and Tommy's cases, it became clear that the courts needed a philosophical analysis of their use of the term. We found that the courts had employed multiple meanings and inconsistent criteria. Worse yet, the arguments in the rulings depended on equivocating between these different meanings of personhood, producing flawed reasoning and rendering the arguments invalid. This was the subject of our amicus curiae brief.

Amicus curiae literally means a "friend of the court." The point of these briefs is to offer information that will help judges reach a just decision. Our brief was written in this spirit, identifying and clarifying the different conceptions of personhood employed in the rulings to date, and explaining what each, when applied rigorously and consistently, implies for Kiko's and Tommy's cases.

One of the strengths of the Common Law is that although it depends heavily on precedent, it also allows judges to exercise their judgment when truly unprecedented cases, like these, come before them (The Regents of the University of California 2010). They can interpret the law in novel ways so long as they are consistent with the intent and spirit of the law and the principles of justice. The Common Law can be flexible in light of changing social values and improved understanding of the relevant empirical facts. Thus, unjust decisions of the past that rested on social prejudices or ignorance need not force the courts to perpetuate the same wrongs. We know that granting Kiko or Tommy habeas corpus relief would be challenging for any judge, requiring considerable moral courage and careful legal reasoning. We hope that our brief and this book offer useful background information and lucid conceptual analyses that can support such a judgment.

PERSONS, ANIMALS, AND RIGHTS

The writ of habeas corpus has a long and storied history in legal cases that challenged the status of individuals not legally recognized as persons, including children, women, and people of color. In the following chapters, we show why the concepts of personhood underlying these decisions should also apply to Kiko and Tommy. As a legal strategy, petitioning for a writ of habeas corpus draws comparisons between chimpanzees and those humans who have historically been denied personhood. We are keenly aware that this is a fraught issue. Comparing African Americans with apes, for example, has long been a staple of racist discourse. Similarly, comparing people with disabilities to nonhuman animals has long been a part of ableist discourse. Some advocates for these groups reject any comparisons between humans and other species, insisting on the radical difference between themselves and animals. Our arguments focus on continuities across humans and animals who have been denied personhood, and so run the risk of being perceived as challenging or weakening the political claims of historically marginalized groups. This is an issue we take very seriously. Our full response to it will emerge over the course of the book as we explore in some detail the connections between notions of personhood that have emerged in the context of human struggles and the plight of Kiko and Tommy. However, it may be useful to provide some preliminary reflections from critical race and disability scholars on the intersections of these different social justice struggles.

Philosopher Syl Ko explains how "almost any good analysis of racism or coloniality usually calls attention to the degree to which racialized folks are *animalized*. That is, we animalize or *dehumanize* certain folks, individually or as groups, thereby justifying their violation" (Ko and Ko 2017: 45). Legal scholar Angela P. Harris notes that the problem with the analogy between slavery and the oppression of animals is that

> it is tone-deaf in a way that covertly exploits the very racism that animal liberationists claim to reject. Precisely because of the close relationship between colored people and animals in the white imagination, the invocation

of the dreaded comparison—the chained slave next to
the chained animal in a sinister visual rhyme—itself
calls out the structures of feeling that have undergirded
racism for so long. The comparison implicitly constructs
a gaze under which slaves and animals appear alike.
This is the sentimental gaze of the privileged Westerner
who "saves" those less fortunate, the voiceless masses
whether human or animal.

(Harris 2009: 25–26)

For Ko, *animality* and *race* are related constructs, concepts that white Europeans have used to separate themselves as both morally and legally superior—one picks out *Homo sapiens* as superior beings, while the other privileges European whiteness. Similarly, Sylvia Wynter argues that the domain of the "human" includes a particular way of being, exemplified by a code of conduct that acts as a conceptual marker for European whiteness as the ideal way of being *Homo sapiens* (Wynter 2015). For this reason, Ko argues, "If animalizing people is problematic, humanizing them is even worse" because it conceives of a person or group as only "*like* they are humans," (Ko and Ko 2017: 21) granting them moral status if and to the extent that they conform to the ideal of European whiteness. Thus, humanization is not merely the act of asserting that one is *Homo sapiens*, but of asserting one's resemblance to white people.

Harris, likewise, notes how the concepts of "animal" and "wild" have been used to undergird the oppression of both African Americans and Native Americans (and, of course, other Indigenous peoples worldwide). "The very notion of what is 'animal' and what is 'wilderness' has been shaped by an European epistemology that has left certain peoples on the wrong side of the paper" (Harris 2009: 32).

We acknowledge that the intersection of race and animality in philosophical (and legal) discussions of personhood is a fraught and painful space, both historically and currently. In particular, many African Americans may understandably bristle at the legal strategy of employing as precedent habeas corpus relief for enslaved Africans (for example, the 1772 British *Somerset v. Stewart* case [Wise 2005]) in an effort to free nonhuman primates. Black scholars have argued, for example, that comparisons between chattel slavery and animal agriculture, when harnessed to gain the support of racially oppressed peoples, can sometimes constitute "emotional blackmail" (McJetters 2015). Such comparisons, moreover, falsely presuppose the post-racialist notion that anti-Black racism has been overcome, thereby allowing "black liberation struggles [to] serve as the positive and negative foil for making a case for the...emancipation of nonhuman beings" (Weheliye 2014:10).

Similar worries have been expressed within the disability community about exploiting their experiences to support the claims of animals (e.g., Carlson 2009). Sunaura Taylor argues persuasively that "[o]ne of the most prevalent lines of argument in defense of animal rights is structured around ableist assumptions about cognitive capacity coupled with a rhetorical instrumentalization of disabled people" (Taylor 2017: 67). Taylor cautions against the strategy—common among animal rights advocates—of making comparisons between animals and the intellectually disabled since they inevitably "miss the more important point that a focus on specific human and neurotypical 'morally relevant abilities' harms both populations" (Taylor 2017: 69).

As a group, we are united in a commitment to eradicating practices of oppression wherever we encounter them. Like Ko, we reject the notion that animality justifies oppression or moral and legal inferiority. Indeed, given the ways that racist, speciesist, and ableist hierarchies are entangled, mutually reinforcing, and energize each other (Kim 2015; Taylor 2016), we believe that challenging the oppression of animals will often be a constructive strategy for other social justice struggles.

Of course, much depends on the details of the arguments. Analogies and comparisons between groups and across species can be drawn in ways that illuminate or obscure fundamental moral and political distinctions. Alice Crary (2018), who similarly rejects the inferiority of animality, argues:

> Given that the notion of ethically charged distinctions between 'lower' and 'higher' animals is in fact still rampant in our culture, and given that this notion has historically been, and still is, integral to rhetoric that contributes to the subjugation of socially vulnerable groups of people—including, not only the cognitively disabled, but also the physically disabled, people identified as non-white, women, the gender-non-conforming, the very old, etc.—we need to be especially careful in tracing out lines of filiation between the lives of animals and the lives of members of human groups who confront systematic forms of bias.
>
> (Crary 2018: 21)

As we develop our arguments, we have tried throughout the book to take these entanglements into account. Our focus is on the case for chimpanzee personhood. When we say that chimpanzees are persons, we aim to *broaden* the concept of person to include chimpanzees without excluding other humans and other persons. We reject any construction of personhood that fails that test.

THE ARGUMENT

There are three rulings on Kiko's and Tommy's cases that appear to rely upon three distinct notions of personhood. The first ruling, by the Third Department of the Appellate Division of the New York Supreme Court, recognizes the "great flexibility and vague scope" of habeas corpus (*People ex rel. Nonhuman Rights Project, Inc. v. Lavery* [2014]: 3), but maintains that it has only ever been applied to members of our species, and this is the clear intent of the law. That is the first conception of personhood – a species membership conception – used by the courts. They also maintain that historically this right and, by implication, all legal rights have been connected to "the imposition of societal obligations and duties" that stem from "principles of social contract" (*People ex rel. Nonhuman Rights Project, Inc. v. Lavery* [2014]: 3). This subtly shifts from a view that membership in our species confers personhood and legal rights to a social contract view, the second conception of personhood the courts used. Finally, there is some suggestion that membership in the human community is ultimately what confers personhood and legal rights, as clearly expressed in the ruling by the First Department of the Appellate Division of the New York Supreme Court. It recognizes that not all humans can bear duties, but maintains that they are nonetheless persons because they are "members of the human community" (*Matter of Nonhuman Rights Project, Inc. v Lavery* [2017]: 6), thus introducing a third distinct conception of personhood.[2]

Each of these different conceptions supports different reasoning regarding personhood. The first, species membership, is untenable due to its arbitrary character. The other two, when properly understood, entail that Kiko and Tommy can qualify as persons. There is yet a fourth conception of personhood, a capacities conception that the NhRP has endorsed, which also entails that chimpanzees are persons. On these grounds we agree with the NhRP that it is unjust to deny Kiko and Tommy habeas corpus relief.

In the next four chapters we explore these four conceptions of personhood, revisiting the arguments presented in the amicus curiae brief. In the final chapter, we summarize the argument of the book, and explore with caution a third possible category for nonhumans, that of sentient being. We conclude by looking at some implications for animals beyond chimpanzees. We include here background and detail we couldn't provide in the original brief due to limits on its length. Although we are now a group of thirteen rather than seventeen authors, the basic structure and character of the argument remain the same. Our goal is not to provide a positive account of personhood. Instead we describe and assess the three conceptions of personhood that can be found in the court decisions and a fourth introduced by the NhRP. Each of these conceptions reflects a view that can be found in the public imagination and in the philosophical and legal literature. Our aim is to explain these conceptions, evaluate their plausibility, and elucidate what each implies for the following question before the courts. Are Kiko and Tommy persons with a right to habeas corpus relief?

NOTES

1 A complete history of these cases is available at the NhRP's website: www.nonhumanrights.org/litigation/.
2 The ruling by the Fourth Department of the Appellate Division of the New York Supreme Court denied the writ of habeas corpus on the grounds that the NhRP was not seeking to release Kiko and Tommy from captivity, but rather to move them to a different captive location (*Matter of Nonhuman Rights Project, Inc. v Presti* [2015]). This ruling, relied upon by the First Department (*Matter of Nonhuman Rights Project, Inc. v Lavery* [2017]), is addressed in Chapter 5, where the importance of an appropriate sanctuary environment is explained.

REFERENCES

Barlow, R. (2017). "Nonhuman Rights: Is it Time to Unlock the Cage?" *Bostonia*. [Online] www.bu.edu/bostonia/summer17/nonhuman-rights-project/ [Accessed June 30, 2018].

Carlson, L. (2009). *The Faces of Intellectual Disability: Philosophical Reflections*. Bloomington: Indiana University Press.

Chan, A. and Harris, J. (2011). "Human animals and nonhuman persons," in T.L. Beauchamp and R.G. Frey (eds.) *The Oxford Handbook of Animal Ethics*, Oxford: Oxford University Press.

Crary, A. (2018). "The horrific history of comparisons between cognitive disability and animality (and how to move past it)," in L. Gruen and F.P. Rapsey (eds.) *Animaladies*. New York: Bloomsbury Press.

Grimm, D. (2018). "U.S. chimp retirement gains momentum, as famed pair enters sanctuary," *Science*. [Online] www.sciencemag.org/news/2018/03/us-chimp-retirement-gains-momentum-famed-pair-enters-sanctuary [Accessed 3 Apr. 2018].

Harris, A.P. (2009). "Should people of color support animal rights?" *Journal of Animal Law*, 5: 15–32.

Kim, C.J. (2015). *Dangerous Crossings: Race, Species, and Nature in a Multicultural Age*. Cambridge: Cambridge University Press.

Ko, A. and Ko, S. (2017). *Aphro-ism: Essays on Pop Culture, Feminism, and Black Veganism from Two Sisters*. Herndon, VA: Lantern Books.

Matter of Nonhuman Rights Project, Inc. v Presti [2015] NY Slip Op 00085.

Matter of Nonhuman Rights Project, Inc. v Lavery [2017] NY Slip Op 04574.

McJetters, C.S. (2015). "Animal rights and the language of slavery," [Blog] *Striving with Systems*. [Online] https://strivingwithsystems.com/2015/12/27/animal-rights-and-the-language-of-slavery/ [Accessed 3 Apr. 2018].

People ex rel. Nonhuman Rights Project, Inc. v Lavery [2014] NY Slip Op 08531.

Taylor, S. (2017). *Beasts of Burden: Animal and Disability Liberation*. New York: The New Press.

The Regents of the University of California. (2010). "The common law and civil law traditions," *The Robbins Collection*, University of California, Berkeley, School of Law. [Online] www.law.berkeley.edu/library/robbins/CommonLawCivilLawTraditions.html [Accessed 3 Apr. 2018].

Weheliye, A.G. (2014). *Habeas Viscus: Racializing Assemblages, Biopolitics, and Black Feminist Theories of the Human*. Durham: Duke University Press.

Wise, S.M. (2005). *Though the Heavens may Fall: The Landmark Trial that Led to the End of Human Slavery*. Cambridge, MA: DaCapo Press.

Wynter, S. (2015). *No Humans Involved*. Hudson, NY: Studio Hudson & Moor's Head Press.

The species membership conception

Two

The Third Department notes that in the past habeas corpus has only been granted to members of our species, Homo sapiens, and relies on this precedent for enforcing a species conception of personhood that excludes Kiko and Tommy (*People ex rel. Nonhuman Rights Project, Inc. v. Lavery* [2014]: 3). The idea is that all and only members of the human species are recognized as persons by the law, and exceptions can be justified solely on the basis of some—currently unspecified—relation to members of our species, which is why in some legal contexts corporations can be treated as persons. If this is right, the courts are using a biological classification to determine the proper scope of legal rights and protections.

We reject the use of biological classifications in general, and the species *Homo sapiens* in particular, as a good basis for determining personhood. We begin by pointing out that current thinking has abandoned the idea that there are essences to species, or that species are natural kinds; it follows that there is no "human nature" upon which to ground personhood. Despite what we now know, in the past reliance on biological classifications for moral standing resulted in deeply troubled practices. The same biological essentialism that underlies the idea that there are morally relevant characteristics shared by all and only members of our species has been used to support the hierarchical classification of kinds *within* our species. Indeed, the role of biology as a discipline in rationalizing discrimination against various human groups stands as a cautionary tale of the ways in which social prejudices can infiltrate the sciences and gain further social traction from that alliance.

We argue in this chapter that *taxonomy* alone provides insufficient grounds for determining personhood, but we do contend that *biology* is relevant to figuring out who is a person. As we show in Chapter 5, the life sciences are crucial for identifying which creatures have the capacities that are often thought to be relevant for personhood. Thinking that capacities are relevant to the definition of personhood is a pretty plausible view—it is the one the NhRP defends. Thinking that the taxonomies employed by biologists can define persons is not.

SPECIES MEMBERSHIP, ESSENTIALISM, AND NATURAL KINDS

The fundamental problem with the Species Membership approach is that it is trying to make a biological category do normative work. The concept of 'personhood,' with all its moral and legal weight, is not a biological one, and it cannot be meaningfully derived from the taxonomic category, *Homo sapiens*. Species is only one level of biological classification that reflects what is sometimes called the "Tree of Life." This is a nested hierarchy showing the comparative relationships of all organisms. Humans, members of the species *Homo sapiens*, and chimpanzees (*Pan troglodytes*) are not only different species but have been placed in different genera (*Homo* and *Pan*, respectively). However, they belong to the same family, *Hominidae*, and so share every other more general classification (order: Primates; class: Mammalia; etc.).[1]

Despite its troubled history and contrary to the evidence, the idea that Homo sapiens has a unique, universal, and stable essence has been remarkably resilient. On this *essentialist* view, we can specify the properties that are both necessary and sufficient to identify any individual organism as a member of the species. Contemporary philosophers, biologists, and other natural scientists have suggested that this approach is antiquated, and, indeed, historians have pointed out that it has been controversial throughout its history (Lovejoy [1936] 2001; McOuat 2009). Biology in particular has provided crucial evidence against essentialist approaches to species classification; since Homo sapiens is a category of species, essentialism fails here as well.

Contemporary philosophers of biology often associate essentialism with natural kinds. Over the centuries, many people have thought that species are natural kinds. Since natural kinds effectively "carve nature at its joints" (Plato [360 B.C.E.] 1925: II 265e), they are thought to be especially important in the sciences. However, whether such natural divisions are neat and fundamental or rough and contingent has been up for debate, and the meaning of "natural kind" has shifted along with the metaphysical assumptions that undergird it. Although specialists continue to debate the subject, contemporary philosophers typically think of natural kinds as groupings of things that are objective rather than subjective, arbitrary, or artificial. Natural kinds are thought to permit scientists to draw inferences about how members of various kinds participate in natural laws. Whereas gold and silver are both natural kinds, "the set of all things weighing over fifty pounds" is not. This view takes natural kinds to have certain essential properties that are necessary and sufficient for identifying individuals as members of a kind. The same biological evidence that undermines the idea of species essences suggests that species are not natural kinds (in this sense) either.

It is tempting to suppose that, even if biological taxonomy cannot do the moral work by itself, when one really understands the types of beings that are grouped into the taxon, *Homo sapiens*, we will recognize a human nature that has a unique moral character and deserves a special legal status. But this just ignores the biology. If there are no essential properties enjoyed by all and only members of our species, then there is no underlying, biologically robust, and universal human nature upon which moral and legal rights can rest. Moreover, any attempt to specify such properties either includes members of other species or risks leaving out a considerable number of humans—often the most vulnerable in our society.

THE TROUBLED HISTORY OF BIOLOGICAL TAXONOMY

In this section, we trace a thread through the history of biological taxonomy, focusing on entanglements between essentialist definitions of species (and *Homo sapiens* in particular) and some of the ideologically driven, socio-political applications for which these have been co-opted. Although our discussion focuses primarily on race and sex essentialisms, we would be remiss to ignore the parallels to ableism and to the discrimination and harm suffered by people with disabilities. Disability has historically been, and continues to be, deeply entangled with racist and sexist ideologies (Boster 2013; Taylor 2017).

In the scientific traditions of Europe, the idea that the organic world could be sorted into natural kinds can be traced back to Aristotle. At the same time, Aristotle noticed the continuity of organic life and articulated a vague "hierarchy of being" into which organisms could be sorted. This came to be called the Great Chain of Being (*Scala Naturae*) (Lovejoy [1936] 2001; Archibald 2014)—an idea taken up by the neo-Platonists in the early Christian era. In this image of Creation, God is at the top just above the angels and archangels, while Man (and we do mean '*man*') stands below, atop a hierarchical ordering of Earthly beings. We can still see this thinking in ordinary ways of talking about higher and lower organisms—a bad habit that sneaks into contemporary scientific papers from time to time (Rigato and Minelli 2013).

In the 16th and 17th centuries, as the study of biology became increasingly systematized, the challenges of classification became particularly acute.[2] As Europeans conquered and colonized new lands, biologists acquired more and more specimens, but they did not have a unified and rigorous method for naming or classifying them. Carl Linnaeus brought order to this chaos in the 1730s by introducing the practice of binomial naming, and the basic structure for classifying species within nested hierarchies that scientists still employ today. His system grouped organisms on the basis of physical similarities and differences, allowing biologists to specify the characteristics that made each organism a member of its kind (Daston and Galison 2007; Müller-Wille 2008). The elegance and power of the Linnaean system, when mixed with contemporaneous ideas about natural law and scientific methods, made it prone to essentialist interpretations.

Linnaeus's work did not sit in a cultural vacuum but reflected and ultimately helped to reify contemporary ideas about racial and sexual differences. By the early 18th century, Euro-American slavery had been thoroughly racialized, driven by the political economy of the Atlantic slave trade (Roberts 2016).

The laws of coverture, determining the secondary status of women, were also firmly ensconced in the Common Law (Stretton and Kesserling 2013).³ Biological classifications were used to *naturalize* these (among other) inequitable laws and practices. Sex was a central organizing feature of Linnaeus's work, and although he does not appear to have had essentialist views of race himself, he identified four varieties of human that were differentiated by skin color and geographical location (Müller-Wille 2015). Over the Modern period, whether nonwhites and women were legally considered persons, part persons, parts of other persons, or nonpersons was a slippery and changeable question; nonetheless, members of these groups were systematically denied many of the most basic rights that were supposedly the birthright of persons.

The juxtaposition of Euro-American chattel slavery with the growth of liberal and democratic principles identifying all men (and sometimes even women) as equal is baffling. Indeed, Charles Mills claims that to explain it one needs a robust theory of ignorance that elucidates the role of "not knowing" in white supremacy (Mills 1997). The biological sciences often played an important role, if not in the origins of the racist and sexist practices and laws that characterized modern Europe and its colonies, then in subsequent justification and reification of these injustices which made them so resistant to change. To modify somewhat Londa Schiebinger's diagnosis of the situation, if non-Europeans and women "were not to be given rights in the newly imagined democratic orders, the nature of [these humans] had to be investigated and shown to be unworthy" (Schiebinger 1996: 170).

By the turn of the 19th century, a variety of research programs into human differences were underway. They assumed that there was an ideal human form of mind and body, free from imperfections, and those who failed to meet this ideal were deficient. Although presented as value-neutral, these research programs presupposed the existence of fundamental differences between the races (Gould 1996) or sexes (Schiebinger 1989) that rendered members of an ever shifting set of racial groups, or women, to be naturally unsuitable to fully participate in political and economic life by virtue of their biological characteristics. As the century progressed, innovations in mathematics offered new ways of doing science. Statistics allowed scientists to investigate social ills and devise political remedies in radical new ways. In some respects, these methods offered an alternative to essentialist thinking; in others, they recapitulated it. All too often, scientifically-minded social reformers focused their efforts on the *average* person, leaving no place in their idealized vision of society for people who deviated too much from the norm (Davis 2006). Thus, people with disabilities, including people with mental health issues, and others deemed socially undesirable were increasingly pathologized and thought of as problems for science to solve (Kevles 1985). The essentialist presuppositions that informed much of this work implied that the putative deficiencies in members of these groups were inevitable. Although it was acknowledged that education and "right upbringing" might alleviate these deeply entrenched biological deficiencies to a modest extent, the divergence of members of these groups from the *normative* human justified their subordination and exclusion from (or marginal access to) political and economic activities.

The hierarchy of the Great Chain of Being played a particular role in racism, invariably placing Europeans at the top and typically placing Africans at the bottom as the link between the rest of humanity and the apes (Gould 1996). As Claire Kim (2015) notes, this places efforts to gain legal rights for nonhuman animals at a "dangerous crossing" with progressive anti-racist movements. The use of legal precedents granting basic rights to people of color rhetorically reasserts the continuous hierarchy of the Great Chain of Being that placed nonwhites next to the apes, and Africans closest of all. That this ordering simply makes no sense in terms of contemporary evolutionary biology is cold comfort when it still seems to inform institutionalized practices of white supremacy (Alexander 2012), and when it still slips into contemporary scientific research programs (see Gould 1996 and Gannett 2004 for critiques).

Early theories of evolution subtly shifted from the Great Chain of Being to the Tree of Life.[4] The similarities and differences schematized in Linnaean classifications were interpreted as degrees of relatedness between various organisms, lineages chronicling the history of ancestor-descendant relations. Through an evolutionary lens, the nested hierarchy of classification transformed into a kind of family tree. The great insight of Charles Darwin (a century after Linnaeus) was that the differences between species did not reflect discrete categories, but instead were the product of a gradual process of natural selection. Darwin (1859) emphasized the diversity found within organic populations, pointing out that the heritable variations that tend to help individuals survive and reproduce in their natural environments become more common in subsequent generations. This results in a slow accumulation of changes, producing distinct varieties within a population and, eventually, a new species. Humans and chimpanzees share a common ancestor. Through various evolutionary processes, including natural selection, this ancestral population diverged and evolved to exhibit the characteristics the descendant populations have today. Darwin's emphasis on variety exploded the notion that the members of biological groups *necessarily* had either hard borders or shared a set of essential characteristics.

Darwin also expunged the progressivism from evolution. Rather than lineages moving toward perfection, natural selection produced populations that adapted to their environments—organisms well-suited for the particular niches they inhabited. The idea of some contemporary populations being more evolved than others in some absolute sense is incoherent from a Darwinian standpoint. Although the morphology of a contemporary species may have changed little, if at all, from its ancestor millions of years ago, it is no less evolved for that. Indeed, this stability suggests it is extremely well-adapted to its environment, and selection pressures have stopped it from drifting into multiple divergent forms. Throughout the history of life on this planet, different types of organisms evolved at different times, producing new niches in which other organisms might thrive, but this is a far cry from a hierarchical ordering or some inevitable march towards perfection (Gould 1998).

Nonetheless, in the 19th century and well into the 20th century, evolutionary ideas mixed with various social theories, notions of essences, and views on progress to horrific effect. Eugenics movements throughout Europe and the Americas were premised on the idea that various groups—people with disabilities (including those with mental health issues), sexual "deviants," the poor, and various racialized groups (including not only the geographical races identified by Linnaeus, but also the Irish, Roma, Jews, and various Indigenous peoples)—were suboptimal variants in the population. While some bemoaned progressive social theories that allowed such groups to thrive and grow in number, others aimed to curtail their social power and limit their numbers through draconian and sometimes deadly social policies. Yet others sought to eliminate particular groups altogether. The "final solution" pursued by the Nazis in the Holocaust might not have been primarily influenced by Social Darwinism, but its magnitude was certainly enabled by Social Darwinists in Europe and North America turning a blind eye to the atrocities.

After World War II, scientists moved away from the kinds of essentialist, typological thinking that was characteristic of scientific racism, and instead toward a population-based approach (Gannett 2001). In 1950 and 1951, UNESCO convened a group of prestigious sociologists, anthropologists, and biologists to author a statement on race. They maintained,

> ...[T]he term 'race' designates a group or population characterized by some concentrations relative as to frequency and distribution of hereditary particles (genes) or physical characteristics, which appear to fluctuate and often disappear in the course of time by reason of geographic and/or cultural isolation.
>
> (UNESCO 1969: 31)

They suggested perhaps doing away with the term 'race' altogether in favor of "ethnic groups" (UNESCO 1969: 31), noting that although populations maintained some distinctiveness through intrabreeding, interbreeding between populations "has been going on from the earliest times" (UNESCO 1969: 33). This was in keeping with the later discovery that the vast majority of genetic variation (85%) is found within populations (Lewontin 2006). These statements brought the human sciences in line with the "Modern Synthesis" in biology, which united Darwinian evolution, genetics, and population biology into a single theory.

Population thinking gives a very different image of the relationship between human racial or ethnic groups (and, indeed, the diversity of abilities and genders throughout our species) compared to the Great Chain of Being or a genealogical Tree of Life. A coarse-grained view of the history of racial diversity would look more like a *braided river*, with smaller populations breaking off from a larger one only to join back again in the future. As noted by the UNESCO group, at a fine-grained view, not even this structure is strictly true. Biologists call this interbreeding "gene flow" and recognize that, although it is often thought of as what binds the members of a species together, it occasionally happens across species through a process known as *hybridization*. They typically ignore the fine-grained analysis because it is too complicated, preferring instead the coarse-grained view. Perhaps the best image for human racial diversity is a *fabric* with each population or ethnic group a *thread connected to every other*, if only indirectly (Gannett 2004). As a number of philosophers of biology have noted, one *can* impose a tree structure on the population of contemporary *Homo sapiens* by using contemporary genetic and statistical techniques—computers make this remarkably easy—but it is not clear why one would want to do so. Certainly, it will tell you *something* about the ancestral relationships between humans, but *what* it tells you will depend on the type of genetic data and statistical methods you happen to use, as well as on the number of terminal branches into which you sort your population (Gannett 2004; Spencer 2016).

This history charts a path from a commitment to species essences—the premise that there is a set of properties shared by all and only the members of any given species—to an understanding that all organisms on the planet are related, with each individual being unique in some ways while being just like its relatives in other respects. Running parallel to, and intertwining with, this scientific history of discovery and classification is a political history of oppression. Essentialist biological taxonomies within the human species were informed by preceding practices of colonialism and enslavement, and were developed in ways that helped to justify the continuation of practices that violated people's basic rights, relegating the majority of humans on the planet to some incoherent mix of both person and thing, or not persons at all. Similar stories can be told about essentialist biological theories of sex (Fausto-Sterling 2000) and of disability (Davis 2006), although these diverge from the story of taxonomic kinds.

The history of human taxonomy reveals how biology has informed the concept of personhood but in pernicious ways that supported ideology and the vested interests of the powerful. It further reveals that personhood is a flexible concept that has changed significantly, even in recent history. Any attempt to base personhood on species membership must come to terms with this social and political history and take pains to draw upon the best contemporary biological science, which refutes the essentialist assumptions that informed pre-Darwinian conceptions of species, race, sex, and disability.

SPECIES IN CONTEMPORARY BIOLOGY

Despite the indications that Darwinian evolution was inconsistent with essentialist taxonomies, many biologists in the 20th century attempted to discover the true essences of species (Ereshefsky 2017). The project was to define the term 'species' so that it picked out biological natural kinds (in the essentialist sense). In effect, the task was to specify an essence for species as a taxonomic level that could be applied to determine the essences of each species and so classify all members of the organic world.

Essentialism about particular species entails that anyone should be able to determine the species of a particular organism by observing that it has certain properties—properties that *all* members of the species have and that are exhibited *only* by members of that species. Since species essentialism requires that this set of properties can effectively *define* the species, it also requires that this definition is *stable* with respect to the place and time in which you are looking. You should be able to identify a particular individual as a member of the species regardless of the context in which you encounter it. If a dog is essentially a highly social and furry quadruped with a wet nose who pants when overheated, then you should be able to say with confidence that Rex is a dog whether you found him on the sidewalk or on another planet.

The gradualism of evolution suggests that there are unlikely to be any species essences to discover. Evolutionary theory reveals that the category of species violates every requirement of species essentialism and therefore cannot be described in terms of natural kinds. There are three central reasons for this.

i. There are no properties that are necessary for an organism to be a member of a particular species—no properties that are universally exhibited by all members of that species across their lifetimes. Instead, there is a substantial amount of *variation* among organisms within a species. In fact, diversity within a species is a precondition for natural selection to even take place.

ii. There is no property (nor set of properties) that is sufficient to identify an organism as a member of a particular species—no traits that are unique to a particular species such that only members of that species can, and do, exhibit it. Instead, there is a great deal of similarity across species because all organisms on the planet are more or less closely related to each other. It is often the case that the more closely related two species are, the more similar they tend to be.[5]

iii. There are no stable properties of species—no traits that can always define a species regardless of when in history one looks. Instead, species *change* over time—they evolve. Even if all members of a species shared some characteristic at one time, this would probably not be true of all their descendants, and it was definitely not true of all their ancestors.

These facts about the process of evolution and the character of living organisms create a fundamental problem for scientists studying the classification of organisms, referred to as the Species Problem. This problem arises when biologists attempt to agree on how to define the concept of taxonomic species. At least twenty different definitions have been suggested (Ereshefsky 2017: Section 1). While among sexual species, interbreeding has often been used to define the boundaries of species groups, this is controversial and leads to its own set of problems and counterexamples from ligers or mules to the eastern coyote.[6]

Evolutionary theory disrupts not only essentialism about species, but also essentialism about all taxonomic categories—the entire Tree of Life (Doolittle and Bapteste 2007), from domains to kingdoms to phyla and down the organizational hierarchy. Evolutionary theory facilitates the grouping aspect of classification, offering a principled criterion for clustering organisms together—shared ancestry—but offers no clear criteria for the level at which to rank them. Whether an ancestral grouping (a "clade") should be considered a variety, subspecies, species, superspecies, subgenus, or genus is an open question.

Given the Species Problem, how can biologists make meaningful use of the concept of species? Contemporary biologists often find it more useful to make generalizations about populations of taxonomically related organisms rather than species. Indeed, this was the lesson that many biologists took from the Modern Synthesis. When it is necessary to discuss a species at all, it is most accurate to consider a species as analogous to a biological individual—a spatiotemporally distinct entity that is born when it splits off from an ancestor (often one that is shared with another species), changes over time in response to the environment, and eventually dies (by going extinct, or splitting into two or more new species—its descendants) (Ereshefsky 2017: Section 2.2).

When understood as a biological classification, it is difficult to see why species, or indeed any other taxonomic category, should bear any moral weight. What counts as a species is a highly contingent decision, and no essential properties are shared by all members of a species to justify special moral status. There are *capacities* that might typically be shared by the members of a particular species that are morally relevant, but then it is the capacities – not species – doing the ethical work. Species membership is at best a heuristic that aids a superficial assessment of moral status.

HOMO SAPIENS AND HUMAN NATURE

The Species Problem applies to humans as well. Like all species, *Homo sapiens* is a product of evolution. As such, three observations about humans are unsurprising.

1. There is a substantial amount of *similarity* between humans and members of other species (especially our closest kin).
2. There is significant *variation* among humans.
3. Characteristics we currently associate with humans (as a species) are subject to *change* over evolutionary time.

It is challenging to identify any set of traits unique to and universal among Homo sapiens. Most either leave out some humans or include other types of organisms. We can arguably come up with a very limited set of characteristics that are unique to Homo sapiens who are alive right now, such as common ancestry, or similar mitochondrial DNA. Critically, these are few and far between. None of them are stable across evolutionary time or robust across contexts, and they all fail to be salient to any moral qualities or to the attribution of personhood (Hull 1986).

We examine (1), (2), and (3) in turn, and consider their implications for the concept of human nature and for any attempt to identify personhood as essentially a human quality.

There is a substantial amount of similarity between humans and members of other species (especially our closest kin)

There have been many efforts to identify characteristics that are uniquely human and separate us definitively from the rest of the animal kingdom. As more is learned about the lives of other animals, traits once believed to be specific to humans are discovered to be shared by members of other species (Andrews 2015). Examples include the use or creation of tools by a wide variety of nonhuman animals (including some primates, cetaceans, corvids, and cephalopods), self-awareness, the development of communication systems, rationality, and autonomy (see Chapter 5).

When two species share a trait, this can be for a variety of reasons, including shared ancestry or convergent evolution.[7] When it comes to closely related species, most differences are differences of degree, not kind. As we discuss in Chapter 5, chimpanzees and humans have many similarities, which is to be expected when two species share a recent common ancestor. But the significant similarity between humans and our closest relatives might be particularly difficult for us to see because our evolutionary proximity has often been overlooked. Chimpanzees are about as closely related to humans as African elephants are to Asian elephants. The lineages that include humans and chimpanzees, respectively, diverged from a common ancestor about four to eight million years ago (Grabowski and Jungers 2017). By comparison, it has been approximately six million years since the existence of the most recent common ancestor of African elephants (including the 'savannah' or 'bush' elephant *Loxodonta africana* and the 'forest' elephant *Loxodonda cyclotis*) and their smaller-eared counterpart, Asian elephants (*Elephas maximus*) (Rohland et al. 2010).[8]

A kind of prejudice might be responsible for our tendency to perceive our species as radically distinct from others in the animal kingdom. A number of theorists (e.g., Diamond 1993: 97; Goodman et al. 1998; Wildeman et al. 2003) have argued that chimpanzees and bonobos (*Pan paniscus*) ought to be reclassified, alongside humans, in the genus *Homo*. This thought is not new. In the 18th century, Linnaeus wrote to a colleague that his reasons for placing Man in a distinct genus had more to do with placating theologians than with the principles of natural history.

> But I ask you and the whole world for a generic differentia between man and ape which conforms to the principles of natural history. I certainly know of none... If I were to call man ape or vice versa, I should bring down all the theologians on my head. But perhaps I should still do it according to the rules of science.[9]

Taxonomic decisions like this are open to revision; this is the lesson of the Species Problem. But here again, we might reasonably worry that politics and prejudice will have as big a role in the ultimate decision as science.

There is significant variation among humans

There have been many efforts to identify characteristics exhibited by every single member of the species *Homo sapiens*. Pinning these down, it is thought, might provide a foundation for our shared human experience, humanity, human nature, and personhood.

Consider three characteristics that seem at first glance to be excellent candidates for a biological foundation of human nature: language, rationality, and morality. Each of these traits invariably falls afoul of the fact that not all humans share it (Hull 1986: 5–6). Linguistic, cognitive, and moral capacities vary greatly among humans. We all spend part of our lives without the ability to linguistically communicate, to make rational inferences, or to engage in moral decision making. Indeed, some of us never acquire these capacities or only acquire them to a limited extent. Thus, none of these candidate traits offers a definitive foundation for identifying an individual as a member of *Homo sapiens* or for grounding human nature or personhood in biological species membership.

Variation is characteristic of all living populations, and we should expect the members of any species to display a wide variety of traits and capacities. Not even the ability of humans to reproduce with one another (interbreeding being one of the favored solutions to the Species Problem) is universally shared across all members of Homo sapiens (e.g., due to choice, isolation, or sterility).[10] Current biomedical efforts to produce human-nonhuman chimeras (thus creating interspecies gene flow) further complicate this criterion for species (Crozier et al. 2016), as does the possibility of creating human-chimpanzee hybrids—a horrific suggestion that has received some recent endorsement (Barash 2018).

Genetics might seem to offer at least a tenuous avenue for identifying species membership. The vast majority of our genome is shared with other organisms (98%–99% with chimpanzees), but even the small part that is unique to humans is not identical in every human.[11] And even if a specific genetic sequence were found in every living human, two further problems arise. First, this is a fragile way to identify Homo sapiens since any modification to, or exception from, this sequence (e.g., by mutation or sexual recombination) could produce a nonhuman indistinguishable from humans in every other way. Second, it is far from clear that this purely genetic difference would be of any moral significance, let alone significant enough to provide a criterion for personhood.

Characteristics we currently associate with humans (as a species) are subject to change over evolutionary time

Our species, like every other, is subject to evolution. For example, the ability to digest lactose in adulthood evolved relatively recently in humans (approximately 7500 years ago in central

Europe), and only in populations where milk-producing livestock were domesticated (Gerbault et al. 2011). No living species is immune to all of the factors (e.g., natural selection, genetic drift, artificial genetic modification, or hybridization) that influence the trajectory of evolution.

Once again we see how a kind of bias might be responsible for our tendency to perceive our species as if it were evolutionarily stable, protected from change over evolutionary time (Tattersall 2017). The fact that none of our closest hominid relatives survived to the present day means that the continuities between humans and other apes might be especially difficult for us to notice.

This raises probing considerations. Suppose Neanderthals (*Homo neanderthalensis*) or *Homo floresiensis* had survived into the current era. They would be another living hominid species. It is an open question whether they would have human rights. The recent suggestion that Neanderthals interbred with modern *Homo sapiens* in Europe might lead scientists to redefine these two species into one. But again, the lesson of the Species Problem is that this would be a contingent decision. There are no natural facts to conclusively decide the matter.

In sum, biology does not support the assertion that personhood can be rationally grounded in being human, because there is no evolutionarily stable and essential property (nor set of properties) universal and unique to the species, *Homo sapiens*, except ones (such as common ancestry) that have no moral bearing. David DeGrazia makes a similar point, concluding, "To single out species as the unique biological basis for moral status is as silly intellectually as it is self-serving for those in whom species prejudice operates strongly" (DeGrazia 2007: 314). If the courts decide to use a biological classification to determine who has legal status, they should explain why species membership should be preferred over, say, genus membership. After all, the lesson of Darwinism is that there are no hard and fast distinctions among any of the categories in the Tree of Life, only a nested organizational hierarchy.

CONCLUSIONS REGARDING SPECIES MEMBERSHIP

We endorse the idea that the biological sciences must inform legal practice, but we do not believe that species membership can rationally be used to determine who is a person or a rights holder. The concept of personhood, with all its moral and legal weight, is not a biological concept and cannot be meaningfully derived from the biological category *Homo sapiens*. Species—including *Homo sapiens*—are not categories identifiable by essences. It follows that there is no method for determining an underlying, biologically robust, unique, and universal human nature upon which moral and legal rights can be thought to rest. Any attempt to specify such a concept either leaves out a considerable number of humans—often the most vulnerable in our society—or includes members of other species.

To many humans, the idea that all and only members of our species could possibly be persons seems compelling; it takes some appreciation of contemporary biology to understand why such a conception is simply not going to work. When we look at the history of biology, we see that the same essentialism and species hierarchy that have been used to place our species morally above all others have played a role in maintaining white supremacy and oppressing women and other persons deemed deficient for various reasons. While biology as a discipline is tainted by this early history, contemporary biology repudiates it. Biologists have discovered that all organisms on the planet are related, each individual being unique in some ways and just like its relatives in other respects.

The NhRP seeks to have Kiko and Tommy classified as persons based on the capacities they share with other persons. If persons are defined as 'beings who possess certain capacities,' and *Homo sapiens* usually possess those capacities, then being a member of *Homo sapiens* can be used to predict with some accuracy that a particular individual will have those capacities and be a person. But it is arbitrary to utilize species membership alone as a condition of personhood, and it fails to satisfy the basic requirement of justice that we treat like cases alike. It picks out a single characteristic as the one that confers rights without providing any reason for thinking it has any relevance to rights.

NOTES

1 *Hominidae* is the taxon that is colloquially referred to as the 'great apes,' but excluding orangutans (Hartwig 2011: 21–22); see Strumpf (2011: 343) for an alternative taxonomy that includes orangutans in *Hominidae*.
2 The study of biology was at the time part of the discipline of 'natural history.' In addition to biology, this discipline also included (but was not limited to) studies which we would now refer to as geology.
3 In his influential *Commentaries on the Laws of England* (1765), William Blackstone explained the basic idea: "By marriage, the husband and wife are one person in law: that is, the very being or legal existence of the woman is suspended during the marriage, or at least is incorporated and consolidated into that of the husband; under whose wing, protection, and *cover*, she performs every thing..." (Blackstone [1765] 2009: 537).
4 Surprisingly, some influential interpretations of the Tree of Life sometimes reiterated the notion of the Great Chain of Being, even while they questioned the unilinear "one true path" of that chain. Branches were often ordered with an eye to the 'top,' to the ascendancy of those 'most advanced'—often the 'white race.' Ernst Haeckel, the great 19th century German biologist and populariser of the image of the great Tree of Life, would often place *Homo sapiens* above '*Homo stupidus*' as if the one 'stood' above the other (Haeckel 1906: 118).

5 It might be objected that for this point and the one above shared ancestry is a property that can identify all and only members of a species. However, this view falls prey to the third consideration as it is rarely possible, except through quite arbitrary stipulation, to determine a strict beginning of a species, never mind the added complexities of hybridization, discussed below. This means identifying the common ancestor all of whose descendants are members of a distinct species is invariably a rough approximation and typically not a particularly informative one.

6 A liger is a hybrid cross of a male lion (*Panthera leo*) and a female tiger (*Panthera tigris*), and the females are capable of reproducing. A mule is a cross between a male donkey (*Equus africanus asinus*) and a female horse (*Equus ferus caballus*). The viability of this particular hybrid is fascinating because the parent species possess a different number of chromosomes—62 in donkeys and 64 in horses; a mule inherits 31 chromosomes from its donkey sire and 32 from its horse dam. The eastern coyote (*Canis latrans var.*) is actually a hybrid of a dog (*Canis familiaris*) and a 'coywolf'; coywolves themselves are hybrids of coyotes (*Canis latrans*) and gray wolves (*Canis lupus*), making the eastern coyote a hybrid of three separate canid species.

7 Convergent evolution occurs when two or more species independently develop the same characteristic because they have adapted to relevantly similar environmental conditions. One commonly cited example of this are the eyes that both octopuses and humans possess despite the fact that their most recent common ancestor was eyeless (Salvini-Plawn and Mayr 1977).

8 The classification of birds, in particular, is notoriously difficult. Consider Bell's vireo (*Vireo bellii*), named after Alexander Graham Bell by ornithologist John James Audubon. This small, North American songbird species consists of two subspecies that differ genetically by as much as 3%, indicating that they diverged from a shared ancestral population somewhere between 1.11 and 2.04 million years ago—a shockingly long time ago for a species with such a short lifespan and generational cycle. Although it has been argued that these two subspecies should be reclassified as two distinct species (Klicka et al. 2016), various complications confound such attempts.

9 Personal correspondence to Johann Georg Gmelin, Uppsala, Sweden, 25 February 1747 (translation from Frängsmyr et al. 1983: 172). Original untranslated letter [Online] http://linnaeus.c18.net/mss_combine/UUB/L-GmelinJG/L0783-a-150-02.jpg.

10 Whether the failure to reproduce is a cultural choice or a biological imposition is irrelevant to speciation and evolution. To illustrate, it is often the case that forced hybridization does not create a new species (see zebra horses and other hybrids discuss in endnote 6). And culturally influenced choice to abstain from breeding with some individual has

been used to identify a new species, even within just a few generations (e.g. the recent speciation event in Galapagos finches due to females finding large finches unattractive). Population bottleneck events and the founder effect—which occur when a sub-population becomes isolated from the larger population, and are a common source of speciation over evolutionary time—is neither a factor of choice nor biological fertility, but of geography. Evolution is undiscriminating when it comes to the reasons why organisms do not breed.
11 See Sarah Richardson's fascinating chapter "Are Men and Women as Different as Humans and Chimpanzees?" for an interesting perspective on this (2013).

REFERENCES

Alexander, M. (2012). *The New Jim Crow: Mass Incarceration in the Age of Colorblindness*, New York: The New Press.

Andrews, K. (2015). *The Animal Mind: An Introduction to the Philosophy of Animal Cognition*. New York: Routledge.

Archibald, D. (2014). *Aristotle's Ladder, Darwin's Tree: The Evolution of Visual Metaphors for Biological Order*, New York: Columbia University.

Barash, D. (2018). "It's time to make human-chimp hybrids: The humanzee is both scientifically possible and morally defensible," *Nautilus* 058: March 8. [Online] http://nautil.us/issue/58/self/its-time-to-make-human_chimp-hybrids [Accessed April 2, 2018].

Blackstone, W. ([1765] 2009). *The Project Gutenberg EBook of Commentaries on the Laws of England*, Book the First, Oxford: Clarendon Press. [Online] www.gutenberg.org/files/30802/30802-h/30802-h.htm [Accessed April 2, 2018].

Boster, D. (2013). *African American Slavery and Disability: Bodies, Property and Power in the Antebellum South, 1800–1860*, New York: Routledge.

Crozier, G.K.D., Fenton, A., Marino, L., Meynell, L., and Peña-Guzmán, D. (2016). "Public comment: Should NIH fund research on human-animal chimeras?" *Hastings Bioethics Forum*, August 30. [Online] www.thehastingscenter.org/challenges-nih-policy-human-animal-chimera-research/ [Accessed April 2, 2018].

Darwin, C. (1859). *On the Origin of Species by means of Natural Selection, or the Preservation of Favoured Races in the Struggle for Life*, London: John Murray. [Online] www.gutenberg.org/ebooks/1228 [Accessed April 2, 2018].

Daston, L. and Galison, P. (2007). *Objectivity*, Cambridge: MIT Press.

Davis, L.J. (2006). "Constructing normalcy: The bell curve, the novel, and the invention of the disabled body in the nineteenth century," in L.J. Davis (ed.) *The Disability Studies Reader*, 2nd edn, New York: Routledge, pp. 3–16.

DeGrazia, D. (2007). "Human-animal chimeras: Human dignity, moral status, and species prejudice," *Metaphilosophy*, 38 (2–3): 309–329.

Diamond, J. (1993). "The third chimpanzee," in P. Cavalieri and P. Singer (eds.) *The Great Ape Project: Equality Beyond Humanity*, New York: St. Martin's Press, pp. 88–101.

Doolittle, W.F. and Bapteste, E. (2007). "Pattern pluralism and the tree of life hypothesis," *PNAS*, 104 (7): 2043–2049.

Ereshefsky, M. (2017). "Species," in E.N. Zalta (ed.) *The Stanford Encyclopedia of Philosophy*, Fall 2017 edn, [Online] http://plato.stanford.edu/archives/fall2017/entries/species/ [Accessed April 2, 2018].

Fausto-Sterling, A. (2000). *Sexing the Body: Gender Politics and the Construction of Sexuality*, New York: Basic Books.

Frängsmyr, T., Lindroth, S., Eriksson, G., and Broberg, G. (1983). *Linnaeus, the Man and His Work*. Berkeley and Los Angeles: University of California Press.

Gannett, L. (2001). "Racism and human genome diversity research: The ethical limits of 'population thinking,'" *Philosophy of Science*, 68 (3): S479–S492.

Gannett, L. (2004). "The biological reification of race," *British Journal for the Philosophy of Science*, 55 (2): 323–345.

Gerbault, P., Liebert, A., Itan, Y., Powell, A., Currat, M., Burger, J., Swallow, D.M., and Thomas, M.G. (2011). "Evolution of lactase persistence: An example of human niche construction," *Philosophical Transactions of the Royal Society of London: Series B, Biological Sciences*, 366 (1566): 863–877.

Goodman, M., Porter, C.A., Czelusniak, J., Page, S.L., Schneider, H., Shoshani, J., Gunnell, G., and Groves, C.P. (1998). "Toward a phylogenetic classification of primates based on DNA evidence complemented by fossil evidence," *Molecular Phylogenetics and Evolution*, 9 (3): 585–598.

Gould, S.J. (1996). *The Mismeasure of Man*, 2nd edn, New York: W.W. Norton.

Gould, S.J. (1998). "On replacing the idea of progress with an operational notion of directionality," in D.L. Hull and M. Ruse (eds.), *The Philosophy of Biology*, Oxford: Oxford University Press, pp. 650–660.

Grabowski, M. and Jungers, W.L. (2017). "Evidence of a chimpanzee-sized ancestor of humans but a gibbon-sized ancestor of apes," *Nature Communications*, 8: 880. [Online] www.nature.com/articles/s41467-017-00997-4 [Accessed April 2, 2018].

Hartwig, W. (2011). "Primate evolution and taxonomy," in C. Campbell, A. Fuentes, K. MacKinnon, S. Bearder, and R. Stumpf (eds.) *Primates in Perspective*, 2nd edn, New York: Oxford University Press, pp. 19–31.

Haeckel, E. (1906). *The Last Words on Evolution: A Popular Retrospect and Summary*, 2nd edn, trans. J. McCabe, London: A. Owen and Co. [Online] www.gutenberg.org/ebooks/53639 [Accessed April 2, 2018].

Hull, D. (1986). "On human nature," *PSA 1986: Proceedings of the Biennial Meeting of the Philosophy of Science Association*, vol. 2, Symposia and Invited Papers, Chicago: University of Chicago Press, pp. 3–13.

Kevles, D. (1985). *In the Name of Eugenics: Genetics and the Uses of Human Heredity*, New York: Alfred A. Knopf, Inc.

Kim, C.J. (2015). *Dangerous Crossing: Race, Species, and Nature in a Multicultural Age*, Cambridge: Cambridge University Press.

Klicka, L., Kus, B.E., Title, P.O., and Burns, K.J. (2016). "Conservation genomics reveals multiple evolutionary units within Bell's vireo (*Vireo bellii*)," *Conservation Genetics*, 17: 455–471.

Lewontin, R.C. (2006). "Confusions about human races," in *Is Race "Real"?: A Web Forum Organized by the Social Science Research Council*, June 7. [Online] http://raceandgenomics.ssrc.org/Lewontin/ [Accessed April 2, 2018].

Linnaeus, C. (1735). *Systema Naturæ per regna tria naturæ, secundum classes, ordines, genera, species, cum characteribus, differentiis, synonymis, locis*, (translates from Latin as *System of nature through the three kingdoms of nature, according to classes, orders, genera and species, with characters, differences, synonyms, places*). [Online] www.biodiversitylibrary.org/item/15373- page/2/mode/1up [Accessed April 2, 2018].

Lovejoy, A.O. ([1936] 2001). *The Great Chain of Being: A Study in the History of an Idea*, Reprint, Cambridge MA: Harvard University Press.

McOuat, G. (2009). "The origins of 'natural kinds': Keeping 'essentialism' at bay in the age of reform," *Intellectual History Review*, 19 (2): 211–230.

Mills, C.W. (1997). *The Racial Contract*. Ithaca: Cornell University Press.

Müller-Wille, S. (2008). "Linnaeus, Carl," in N. Koertge (ed.) *Complete Dictionary of Scientific Biography*, vol. 22, Detroit: Charles Scribner's Sons, pp. 314–318.

Müller-Wille, S. (2015). "Linnaeus and the four corners of the world," in K.A. Coles, R. Bauer, Z. Nunes, and C.L. Peterson (eds.) *The Cultural Politics of Blood, 1500–1900*. New York: Palgrave Macmillan, pp. 191–209.

Pellegrin, P. (1982). *Aristotle's Classification of Animals: Biology and the Conceptual Unity of the Aristotelian Corpus*, trans. A. Preus, Berkeley: University of California Press.

People ex rel. Nonhuman Rights Project, Inc. v Lavery [2014] NY Slip Op 08531.

Plato ([360 B.C.E.] 1925). "Phaedrus," 360 B.C.E., in *Plato in Twelve Volumes*, Vol. 9, trans. H.N. Fowler, Cambridge, MA: Harvard University Press: section II 265e. [Online] www.perseus.tufts.edu/hopper/text?doc=plat.+phaedrus+265e [Accessed April 2, 2018].

Richardson, S.S. (2013). *Sex Itself: The Search for Male and Female in the Human Genome*, Chicago: The University of Chicago Press.

Rigato, E. and Minelli, A. (2013). "The great chain of being is still here," *Evolution: Education and Outreach*, 6 (18): 1–6.

Roberts, J. (2016). "Race and the origins of plantation slavery," *Oxford Research Encyclopedia of American History*, March 3. [Online] http://americanhistory.oxfordre.com/view/10.1093/acrefore/9780199329175.001.0001/acrefore-9780199329175-e-268 [Accessed April 2, 2018].

Rohland, N., Reich, D., Mallick, S., Meyer, M., Green, R.E., Georgiadis, N.J., Roca, A.L., and Hofreiter, M. (2010). "Genomic DNA sequences from mastodon and woolly mammoth reveal deep speciation of forest and savanna elephants," *PLOS Biology*, 8 (12). [Online] http://journals.plos.org/plosbiology/article?id=10.1371/journal.pbio.1000564 [Accessed April 2, 2018].

Salvini-Plawn, L.V. and Mayr, E. (1977). "On the evolution of photoreceptors and eyes," *Evolutionary Biology*, 10: 207–263. [Online] www.pnas.org/content/115/11/E2566 [Accessed April 2, 2018].

Schiebinger, L. (1989). *The Mind Has No Sex? Women in the Origins of Modern Science*, Cambridge: Harvard University Press.

Schiebinger, L. (1996). "Gender and natural history," in N. Jardine, J.A. Secord, and E.C. Spary (eds.) *Cultures of Natural History*, Cambridge: Cambridge University Press, pp. 163–177.

Spencer, Q. (2016). "Do humans have continental populations?" *Philosophy of Science*, 83 (5): 791–802.

Stretton, T. and Kesselring, K.J. (2013). *Married Women and the Law: Coverture in England and the Common Law World*, Montreal: McGill-Queen's University Press.

Strumpf, R. (2011). "Chimpanzees and bonobos," in C. Campbell, A. Fuentes, K. MacKinnon, S. Bearder, and R. Stumpf (eds.) *Primates in Perspective*, 2nd edn, New York: Oxford University Press, pp. 321–344.

Tattersall, I. (2017). "History and reality of the genus 'Homo': What is it and why do we think so?" *Mètode Science Studies Journal*: March 25. [Online] https://ojs.uv.es/index.php/Metode/article/view/9111/10900 [Accessed April 2, 2018].

Taylor, S. (2017). *Beasts of Burden: Animal and Disability Liberation*, New York: The New Press.

UNESCO (1969). "Statement on race," *Four Statements on the Race Question*, United Nations Educational, Scientific and Cultural Organization, Paris, July 1950.

Wildeman, D.E., Uddin, M., Liu, G., Grossman, L.I., and Goodman, M. (2003). "Implications of natural selection in shaping 99.4% nonsynonymous DNA identity between humans and chimpanzees: Enlarging genus homo," *PNAS*, 100 (12): 7181–7188.

The social contract conception
Three

A social contract theory holds that at least some features of our moral and legal relationships are products of a social contract. Our aim in this chapter is to argue that on plausible interpretations of social contract theory, nonhuman animals such as Kiko and Tommy can be persons with rights.

There are two main kinds of social contract theory. The first, *contractarianism*, descends from the social contract theory of Thomas Hobbes. Hobbesian contractarianism holds that contractors are *instrumentally* rational; that is, they are self-interested and capable of understanding how their own interests are furthered and protected by mutual cooperation. We will discuss below the classic contractarian thought of Hobbes, Locke, and Rousseau and argue that plausible interpretations of contractarianism are compatible with the idea of nonhuman personhood and rights.

The second main kind of social contract theory, *contractualism*, descends from Immanuel Kant's social contract theory. On this view, contractors are rational, partly self-interested, and partly other-interested, and so they are motivated to enter social contracts not only for their own sake but also for the sake of others. We will discuss below the contractualist political philosophy of John Rawls as well as of contemporary interpreters such as Mark Rowlands and show that plausible interpretations of contractualism are compatible with, and can even support, the idea of nonhuman personhood and rights.

We argue that Kiko and Tommy can be persons with rights on plausible interpretations of social contract theory. The courts, in ruling on Kiko's and Tommy's cases, have tried – we think unsuccessfully – to argue otherwise. The Third Department provides a representative claim:

> Reciprocity between rights and responsibilities stems from principles of social contract, which inspired the ideals of freedom and democracy at the core of our system of government.... Under this view, society extends rights in exchange for an express or implied agreement (a social contract) from its members to submit to social responsibilities. In other words, 'rights [are] connected to moral agency and the ability to accept societal responsibility in exchange for [those] rights'.
>
> (*People ex rel. Nonhuman Rights Project, Inc. v. Lavery* [2014]: 3)

We will show, however, that this interpretation is based on at least three mistaken assumptions about or interpretations of social contract theory. First, it assumes that the social contract produces persons when, in fact, persons exist prior to the social contract. Second, it assumes that the social contract produces *all* rights when, in fact, some rights exist prior to (and outside of) the social contract. Third, it assumes that one must have duties in order to be a person when, in fact, one can be a person without having duties. Thus, we argue this exclusionary interpretation of social contract theory is not only arbitrary and implausible, it is also incompatible with the classic social contract theories that inspired the U.S. Constitution, as well as with contemporary social contract frameworks that inspire modern human rights discourse.

THE SOCIAL CONTRACT DOES NOT PRODUCE PERSONS

In the philosophies of Thomas Hobbes, John Locke, and Jean-Jacques Rousseau, the social contract does not produce persons. Instead, and at most, it produces citizens (Hobbes 1651; Locke 1689; Rousseau 1762). Persons become citizens who are subject to laws in virtue of their participation in the contract. This means that one does not need to be part of any social contract to be a person.

The Third Department, however, seems to maintain that personhood depends on the social contract, such that we *become* persons only after we enter the social contract. This confuses what the social contract *requires* with what it *produces*. The court effectively has it backwards. It assumes that contracts produce persons when, in reality, it is persons who produce contracts. Conceptually, it makes sense that one must already be a person *before* one enters the social contract because the social contract itself presupposes the existence of persons who will be its creators and eventual signatories. Contracts, therefore, cannot and do not produce persons. They transform persons into citizens. In this framework, citizens are persons who become subject to laws in virtue of participation in the contract.

This distinction between personhood (which the contract requires) and citizenship (which the contract produces) is essential to the contractarian tradition. Locke underscores it when he defines political power, which the social contract institutes, as a relation between a government and its citizens (Locke 1689: 382). Likewise, Rousseau says that the social contract exists not to turn each associate into a person but rather to *protect* "each associate's person and goods," which they possess before the contract (Rousseau 1762: 46). Notably, the U.S. Constitution mentions the term 'persons' fifty-seven times but does not define it. Yet, following the social contract theory that inspired it, it also rightly distinguishes persons from citizens in the 14th Amendment.

The NhRP argues that Kiko and Tommy should be recognized as *persons*, not that they should be recognized as *citizens* (with all the particular rights and responsibilities of citizenship). This distinction is crucial because citizenship rights vary from place to place, whereas personhood rights do not. For example, non-citizens might not have full citizenship rights in the U.S., but they still have full personhood rights including a basic right to bodily liberty.

In the philosophies of Hobbes and Rousseau, with the advent of the social contract, we see the creation of a kind of "artificial man" (the sovereign or Leviathan), a vessel for the collected powers of all its subjects. This artificial man is an abstraction since no real person could be literally composed of the rights and powers of others. This artificial man is the only kind of 'person' a social contract can create.

CONTRACTORS MUST BE PERSONS, BUT PERSONS NEED NOT BE CONTRACTORS

There are different possible interpretations of what it takes to be a contractor, some more demanding than others. As understood by classic contractarians like Hobbes, Rousseau, and Locke, a contractor must have *rationality*, understood as the capacity to engage in means-end reasoning, and *autonomy*, understood as the capacity for self-rule, or the ability to have one's actions directed by oneself in some sense. Without rationality, individuals would not be capable of entering into contracts for their own benefit. Without autonomy, they would not have the authority over themselves that is needed to render the contract legitimate. Therefore, the principles of social contract theory maintain that only *rational and autonomous persons* can voluntarily place themselves under the authority of another, and., thereby, legitimize that authority. Indeed, it is difficult to see how it could be otherwise.

While there may be value in determining whether or not Kiko and Tommy can be contractors according to different versions of social contract theory, this question is ultimately irrelevant to the issue of nonhuman *personhood* given that social contract theory holds that while all contractors must be persons, not all persons must be contractors. Indeed, as we will see, the plausibility of social contract theory depends on the possibility of persons who are not contractors— either because they choose not to contract (e.g., adults who opt for life in the state of nature) or because they cannot contract (e.g., infants and some individuals with cognitive disabilities). Thus, even if Kiko and Tommy cannot (or choose not to) contract, it does not follow that they cannot be persons.

THE SOCIAL CONTRACT DOES NOT PRODUCE ALL RIGHTS

Hobbes, Locke, and Rousseau maintain that all persons have 'natural rights' prior to the existence of the social contract and independently of their willingness or ability to take on social responsibilities (Hobbes 1651; Locke 1689; Rousseau 1762). These natural rights include the right to absolute freedom and liberty. Upon contracting with our fellows, we thus do not suddenly acquire rights but rather give up all or most of our natural rights, sometimes in exchange for civil and legal rights.

The U.S. founding documents reflect this view. For example, the Declaration of Independence says: "We hold these truths to be self-evident, that all men are created equal, that they are endowed by their Creator with certain unalienable Rights, that among these are Life, Liberty and the pursuit of Happiness." Similarly, the U.S. Constitution holds that one of the purposes of the state is to secure (not create) the blessings of liberty:

> We the people, of the United States, in Order to form a more perfect Union, establish Justice, insure domestic Tranquility, provide for the common defence, promote the general Welfare, and secure the Blessings of Liberty to ourselves and our Posterity, do ordain and establish this Constitution for the United States of America.

The Third Department, however, maintains that persons acquire rights the moment they assent to an "express or implied" social contract (*People ex rel. Nonhuman Rights Project, Inc. v. Lavery* [2014]: 3). The social contract, according to this interpretation, is the mechanism whereby persons take up societal duties and responsibilities, receiving rights in exchange. But as before, this is not how the social contract theorists who inspired the U.S. founding documents understand the relationship between rights and contracts. Once again, the Third Department has it backwards.

For Hobbes, the purpose of the social contract is not to acquire rights but to protect us from others exercising their natural rights and liberties against us. Hobbes believes that we have natural rights prior to entering into a contract. These include absolute liberties to do as we see fit to secure our own lives and happiness. For Hobbes, the social contract creates laws, not rights. Indeed, Hobbes believes that in the act of entering a social contract, we give up nearly all of our natural rights save one: the right to life. What we receive in exchange for giving up all these rights are not *new rights* but the security of our persons in the form of the sovereign's protection.

Locke, for his part, believes that we enter society to protect the institution of private property. On this view, individuals in the state of nature create private property by applying their labor to natural resources, giving them *value*. These individuals then gain property rights over these resources, rights they can legitimately claim in the state of nature. For Locke, we agree to a contract to leave the state of nature and form "one body politic under one government" because we have a shared interest in protecting the property rights that we already have. In this transition from the state of nature to the state of civil society, we gain other valuable things too, including laws, executive power to enforce laws, and impartial judges to adjudicate property disputes. But we also lose some rights, most notably our previously held personal rights to protect ourselves by any means necessary and to punish those who transgress against our property.

Finally, like both Hobbes and Locke, Rousseau believes that we have natural rights prior to entering a social contract. For Rousseau, we enter a social contract because we wish to protect the natural rights that we already possess in the state of nature, such as the right to liberty. Rousseau stresses this point when he famously writes, "Man is born free, and everywhere he is in chains" (Rousseau 1762: 41). The idea is not that *others* place us in chains, but rather that *we* place ourselves in chains when we give up our natural rights and freedoms and accept the authority of another. Similarly, in *The Social Contract*, Rousseau stresses the 'natural' origin of many of our rights when he clarifies that if the social contract were to ever dissolve (for whatever reason), each individual would simply be "restored to his original rights and resume his natural liberty" (Rousseau 1762: 50). On this view, the dissolution of civil society would not entail the destruction of all rights but would rather entail the return to a state of nature replete with natural rights.

Thus, according to the social contract theorists whose work both inspires and coheres with the U.S. founding documents, the social contract does not endow persons with natural rights. Through the social contract, we bind ourselves to one another and to the state. The social contract that emerges from this binding is not the source of our natural rights and freedoms. It is merely the justification for the limits civil society places on these rights and freedoms.

PERSONHOOD DOES NOT DEPEND ON BEARING DUTIES AND RESPONSIBILITIES

The First Department claims that "nonhumans lack sufficient responsibility to have any legal standing" (Matter of Nonhuman Rights Project, Inc. v. Lavery [2017]: 5). The Third Department also argues that chimpanzees, unlike human beings, "cannot bear any legal duties, submit to societal responsibilities or be held legally accountable for their actions" (People ex rel. Nonhuman Rights Project, Inc. v. Lavery [2014]: 4), and thus cannot have legal rights. Further, citing Gray, it states that "the legal meaning of a 'person' is 'a subject of legal rights and duties'" (People ex rel. Nonhuman Rights Project, Inc. v. Lavery [2014]: 4).

This claim is worthy of discussion, but first, it is necessary to dispense with the part of the claim that is based on a simple typographical error. The statement that a legal person is one who is a "subject of legal rights *and* duties" is based on a misquote that the NhRP traced to *Black's Law Dictionary*, misquoting John W. Salmond's *Jurisprudence* (1907). The latter actually states that persons are subjects of "legal rights or duties" (Salmond 1907: 275) (This example should be instructive for students or others unclear about how much might hinge on the words we use. The NhRP has been assured by *Black's* that the error will be corrected (Garner 2017:1; Schneider 2017: 1–2; Garner 2017:1).

The NhRP has argued that an entity is a person if they can bear rights or responsibilities. One reason for this broader interpretation is simple: not all persons can be held accountable for their actions and bear societal duties. Infants, children, and those found not guilty by reason of insanity cannot be held accountable and cannot bear societal duties. They are, nonetheless, persons with rights. To be a bearer of rights is thus distinct from being a bearer of duties and responsibilities. To be sure, many persons are both, but one can still be a person *without* being capable of bearing duties. Here again, the courts have it backwards. Having rights and responsibilities presupposes that one is a person. That Kiko and Tommy do not have any legal responsibilities, duties, or rights is the *consequence* of their having been denied the status of persons, not the cause of it.

The upshot here is that one can be a person and have rights whether or not one also has the capacity to bear duties. This idea is plausible and widely accepted and is importantly inclusive of all humans, regardless of their capacities.

SOCIAL CONTRACT THEORY COULD SUPPORT NONHUMAN PERSONHOOD

In this section, we will consider the contractualist theory of political philosopher John Rawls, as well as his contemporary interpreters, such as Mark Rowlands, to show that social contract theory can support the idea of nonhuman personhood.

In *A Theory of Justice* (1971), Rawls advances a social contract theory rooted in the idea of justice as fairness. As a contractualist, Rawls sees social contractors as partly self-interested and partly other-interested, with the result that they are motivated to identify and follow principles that could be rationally justified to others. To help us identify these principles, Rawls encourages us to imagine ourselves in what he calls the "original position," which he defines as "a purely hypothetical situation characterized so as to lead to a certain conception of justice" (Rawls 1971: 11). In this position, our task is to evaluate principles of justice from behind what Rawls calls the "veil of ignorance." That is, our task is to consider what kind of society we want without knowing anything about our own position in that society, including our social or economic status, our race, gender, natural talents, or endowments. Rawls's wager is that the veil of ignorance will help us to be more impartial in our thinking about what kind of society is just. When no one knows what position they will end up occupying, everyone will be motivated to create a fair, equitable, and reasonable social order.

Rawls's political philosophy presupposes that the contractors are, as Rawls puts it, "free and rational persons" who are capable of thinking about principles of justice. But this does not mean that the contractors will create a society that extends justice only to other free and rational persons. After all, behind the veil of ignorance, contractors know that they might not be rational in the relevant sense after the veil is lifted. For example, they might be someone who has not yet developed the capacity for rationality, someone who has lost this capacity, or someone who will never develop this capacity at all. Because the veil prevents contractors from knowing anything about who they might be, the rational thing for them to do is create a society in which all individuals are "treated in accordance with the principles of justice" (Rawls 1971: 441), including those who lack the capacities needed to be contractors.

Rawls's own view about whether contractors would extend the protection of justice to those who lack the capacities needed to be contractors is unclear. In places, he seems to imply that to be a subject of justice, one must have the "two moral powers" of a rational life plan and a sense of justice (Rawls 1971: 505–506). In other places, however, he acknowledges that possession of these two moral powers is sufficient but not necessary for inclusion in the scope of justice. Thus, Rawls's contractualism is compatible with the idea that non-rational beings can be subjects of justice under a social contract (Hartley 2009).

In "Contractarianism and Animal Rights" (1997), Mark Rowlands argues that Rawls's theory of justice can and should extend to nonhuman animals. Central to his theory is Rawls's idea that no one is morally entitled to benefit from undeserved characteristics. For Rawls, examples of undeserved characteristics include race, gender, intelligence, and social and economic status at birth. Drawing from this idea, Rowlands argues that, although Rawls himself is somewhat coy about whether or not we can know our species behind the veil of ignorance, and about whether or not nonhuman animals can be subjects of justice, species is another example of an undeserved characteristic. Like the capacity for rationality, species membership is a natural endowment for which we are not responsible. Rowlands concludes that Rawlsian impartiality requires that we not privilege members of the species, Homo sapiens, when thinking about the requirements of justice.

If this is right, then it follows from Rawlsian contractualism that justice need not, and arguably should not, be restricted to humans. In the same way that non-rational humans can be subjects of justice according to Rawlsian contractualism, nonhuman animals can also be subjects of justice. In both cases the justification is the same. The rational agents who must evaluate principles of justice from behind the veil of ignorance understand that they do not know who they will be once the veil is lifted. They may or may not be human, and they may or may not be rational. Moreover, even if they are rational, they may or may not always remain that way. Therefore, they will be motivated to extend justice to everyone who merits moral and legal consideration, regardless of undeserved characteristics such as race, gender, species membership, capacity for intelligence and rationality, and social and economic standing at birth.

But not everybody accepts that social contract theory can lead to this result. For example, in *Frontiers of Justice* (2006), Martha Nussbaum expresses skepticism that this strategy of extending justice to nonhuman animals will succeed. Nussbaum concludes that we should reject social contract theory altogether, not that we should reject the idea of justice for nonhuman animals. In particular, Nussbaum thinks that social contract theory requires more universality of capacities than we have in the real world. As we've seen, both the Hobbesian contractarians and the Kantian contractualists begin by imagining a situation where contractors form an agreement from a position of equal power. For Hobbes, this is the state of nature where contractors have equal power to harm one another; for Rawls, this is the veil of ignorance which equalizes power by preventing people from knowing what social position they will occupy. In Nussbaum's view, this strategy of imagining a contracting situation of equal power is unrealistic and unhelpful. It ignores the depth and pervasiveness of unequal power in the real world, whether amongst humans or between humans and animals. Nussbaum thus invites us to search for a different way to extend justice to animals. This different way, which is rooted in capacities, also lends support to the view that nonhuman animals can be persons (as we show in Chapter 5).

We are not convinced that social contract theory does, in fact, require more in the way of shared capacities than we currently have in the real world to make sense as a framework for thinking about justice. If, following Rowlands, we view the social contract as an invitation to abstract away from undeserved characteristics and think impartially about justice, then we have reason to claim exactly the opposite—namely, that social contract theory helps us promote justice by envisioning a world not mired in social inequality, a world that can act as a regulative principle for contemporary struggles against injustice. Either way, Nussbaum's critique of social contract theory seems to boil down to a worry about its inherent idealism. If we disagree with her in this regard, we can hold on to the view that there is value in social contract approaches to justice. If we agree, then the failure of the theory is partly a function of its inability to accommodate nonhuman animals within the sphere of justice. In this case, we can reject social contract theory in favor of another, more inclusive framework. Regardless of which of these options we select, what matters is that we cannot reasonably accept a social contract theory that denies justice to "non-rational" human or nonhuman animals. On this point, Rowlands and Nussbaum agree. And we agree with them.

OTHER CONTRACTARIAN ARGUMENTS FOR EXCLUDING NONHUMANS

Law professor Richard Cupp, in a series of articles and in an amicus brief submitted to the First Department, offers a peculiar interpretation of the social contract according to which all and only humans can be persons with rights. His basic argument for this conclusion is simply that when we, as a society, decide to extend personhood (and, with it, what Cupp calls "dignity rights") to infants, children, and people with certain "cognitive impairments" (to use Cupp's phrase), we do so on the basis of their humanity, not on the basis of their rationality or autonomy. Thus he maintains that humans who lack rationality or autonomy are still persons, and cites their "humanity" as justification. As Cupp puts it, "Humanity has historically been central to rights" (Cupp 2013: 33).

However, Cupp's claim that only humans can be persons is muddled at best and viciously circular at worst. First, he confuses matters of historical fact with matters of normative justification. The fact that historically we have appealed to the humanity of vulnerable individuals in order to ascribe to them legal personhood does not mean that their humanity is what legitimates this ascription. It could be that we have simply used the term 'humanity' as a proxy for another foundation that may be correlated but not exactly coextensive with it. Second, Cupp's appeal to humanity as the basis for personhood is question begging in the present context, since the relationship between humanity and personhood is precisely what is in question here. The NhRP maintains that Kiko and Tommy are nonhuman persons who have been denied what Cupp calls "dignity rights" (such as the right to bodily liberty) *because* we have mistakenly assumed that humanity is the sole basis of legal personhood. In this context, to say that a particular being is a person *because they are human* presupposes exactly what is legally and philosophically at stake.

The main problem with Cupp's argument, however, is that it is unabashedly speciesist, and therefore arbitrary and inconsistent with justice. Cupp maintains that all and only members of the species *Homo sapiens* can be persons with basic rights because, somehow, all and only members of that species are endowed with the sort of "dignity" that warrants having rights. In Chapter 2 we outline the many conceptual, scientific, and normative problems associated with this biological conception of personhood, including its arbitrariness and incompatibility with principles of justice. Cupp's view amounts to little more than the assertion that "humans are persons because humans are humans."

Cupp's appeal to history, meanwhile, does little to bolster this argument. He asserts that U.S. law has never recognized any nonhuman as a person (except for entities like ships and corporations in which humans have important legal interests, as a mechanism for extending some personal rights). This much is correct. However, he ignores two important facts. First, he ignores the fact that U.S. law has not always recognized *all* humans as persons, as the historical legal exclusion of African Americans, women, and children illustrates. Indeed, in 1874, lawyers for the American Society for the Prevention of Cruelty to Animals filed a writ of habeas corpus on behalf of an abused child, since, at the time, children lacked the status of legal persons and there were no laws in New York preventing their abuse (Markel 2009: 1). Second, Cupp also ignores the legal fact that, although asserting that chimpanzees are persons is indeed novel and without precedent in American law, just as asserting that children are persons was once novel and without precedent, the novelty of the assertion is not by itself sufficient reason to reject it. Indeed, it is one of the proper roles of the courts to adjudicate such novel claims.

Peter Carruthers (1992) constructs a more sophisticated contractualist argument for the idea that all and only humans have rights. However, his argument ultimately fails too and for similar reasons. Carruthers argues that only humans can have rights on the grounds that only humans are rational agents and only rational agents can have rights. Carruthers has a demanding take on the rationality required for contractualism. In his view, being a rational agent requires having beliefs and desires, but also being able to construct long-term plans in the light of those beliefs and desires. In addition, rational agents have an understanding of what it is to act under universalizable rules (e.g., Do not lie), and can think about what it would mean for those rules to be generally adopted (Carruthers 1992). Of course, many humans are not rational in this demanding sense. Yet, Carruthers argues we should treat all humans as having rights anyway, for two reasons that he says do not extend to nonhumans.

First, Carruthers holds there is no sharp boundary between rational and non-rational humans in the real world. If we try to determine whether or not particular humans are rational, "we shall be launched on a slippery slope which may lead to all kinds of barbarisms" (Carruthers 1992: 114). Yet, he claims there is a sharp boundary between humans and nonhumans. Thus, we can safely classify nonhumans as non-rational without the risk of a slippery slope (Carruthers 1992: 115).

Second, Carruthers says that humans care intensely about other humans. If we were to deny rights to "non-rational" humans, "many people would find themselves psychologically incapable of living in compliance" with this rule (Carruthers

1992: 117). Yet, Carruthers argues that humans do not care about nonhumans with the same intensity. Thus, we can deny rights to nonhumans without risking social instability.

Carruthers' position is essentially a version of what in Chapter 4 we call "personhood-by-proxy." As we discuss in more detail there, this kind of argument tends to be problematic empirically as well as normatively. Consider, for instance, the empirical problems with Carruthers' position. It is not clear that classifying humans as "non-rational" leads to a slippery slope while classifying nonhumans as "non-rational" does not. In both cases our psychological associations are likely to be complex. Historically, we have made distinctions within our own species and associations across species. Indeed, part of how humans rationalize the oppression of other humans is by *dehumanizing* them, an act which involves making distinctions within our own species (between the oppressor and the oppressed) and associations across species (between the oppressed and nonhumans thought to lack morally relevant features such as rationality) at one and the same time. The inconsistency of these distinctions and associations is part of what makes oppression so insidious. Likewise, it is not at all clear that we care about all humans as much as Carruthers claims, or that we care about nonhumans as little as he claims. In both cases, our patterns of care tend to be inconsistent, again without any clear implications for what kinds of exclusions will risk social instability.

Now consider the normative problems with Carruthers' position. Carruthers bases his argument on an unacceptably *indirect* account of the moral and legal status of "non-rational" humans. In particular, Carruthers is claiming that we should treat "non-rational" humans as having rights not for their *own* sake but rather for the sake of other, "rational" humans whose well-being depends on social recognition and social stability. But this is decidedly not the main reason we should treat "non-rational" humans as having rights. The main reason we should treat them as having rights is that *they have rights*. In particular, they have morally and legally relevant capacities and relationships, and so they merit moral and legal recognition whether or not others are likely to benefit from them having rights, and whether or not they can be rational in the demanding sense that Carruthers requires. Once we accept this basic truth, we realize Carruthers' *indirect* reason for extending rights to those deemed non-rational by his theory is inadequate.

Some social contract theorists, such as Cupp and Carruthers, deny that contractarianism and contractualism can be compatible with nonhuman personhood and rights. As we have demonstrated, their reasons for doing so do not withstand critical philosophical scrutiny, and in order to exclude all nonhumans and include all humans, they must resort to arbitrary and question-begging claims.

CONCLUSION

We have argued that, on plausible interpretations, social contract theory does not rule out the possibility of nonhuman personhood. It is not only compatible with the idea of nonhuman personhood but can even support it. Granted, some interpret social contract theory in an exclusionary way. But as we have seen, this exclusionary interpretation is arbitrary, implausible, and incompatible with the best, most influential historical and contemporary social contract frameworks. On these frameworks, one can be a person without being a contractor, one can have rights without being a contractor, and one can be a person and have rights without having duties and responsibilities. Moreover, if we view social contract theory as a framework for thinking impartially about justice, then it can support the idea of nonhuman personhood and rights. This does not mean that nonhuman persons like Kiko and Tommy would have all the same rights as humans. The rights being claimed for Kiko and Tommy under the writ of habeas corpus are that they be recognized as persons with rights against unlawful detention. We might minimally describe their rights as a right to bodily liberty, or "life, liberty, and the pursuit of happiness." Such minimal rights would certainly preclude the kind of solitary confinement and captivity in which they live now.

REFERENCES

Carruthers, P. (1992). *The Animals Issue: Moral Theory in Practice*, Cambridge: Cambridge University Press.

Cupp Jr, R.L. (2013). "Children, chimps, and rights: Arguments from marginal cases," *Arizona State Law Journal*, 45 (1): 1–52.

Garner, Bryan. (2017). *Correspondence*. [Online] www.nonhumanrights.org/content/uploads/NhRP-email-to-Bryan-Garner-re-Person-Definition-4.6.17.pdf.

Hartley, C. (2009). "Justice for the disabled: A contractualist approach," *Journal of Social Philosophy*, 40 (1): 17–36.

Hobbes, T. (1651) [1994]. *Leviathan: With selected variants from the Latin edition of 1668*, Indianapolis: Hackett Publishing Company.

Locke, J. (1689) [1988]. *Two Treatises of Government*, Cambridge: Cambridge University Press.

Markel, H. (2009). "Case shined first light on abuse of children," *The New York Times*, 14 December 2009. [Online] www.nytimes.com/2009/12/15/health/15abus.html.

Matter of Nonhuman Rights Project, Inc. v. Lavery (2017 NY Slip Op 04574).

Nussbaum, M.C. (2009). *Frontiers of Justice: Disability, Nationality, Species Membership*, Cambridge, MA: Harvard University Press.

People ex rel. Nonhuman Rights Project, Inc. v Lavery. 124 A.D. 3d 148 (3d Dept. 2014).

Rawls, J. (1971). *A Theory of Justice*, Cambridge, MA: Harvard University Press.

Rousseau, J.J. (1762) [1997]. *Rousseau: The Social Contract and Other Later Political Writings*, Cambridge: Cambridge University Press.

Rowlands, M. (1997). "Contractarianism and animal rights," *Journal of Applied Philosophy*, 14 (3): 235–247.

Salmond, J.W. (1907). *Jurisprudence: Or the Theory of the Law*, London: Stevens and Haynes.

Schneider, Kevin. (2017). *Correspondence*. [Online] www.nonhumanrights.org/content/uploads/Letter-to-Blacks-re-Def.-of-Person-4.6.17-ks-2.pdf.

The community membership conception
Four

As we discussed in Chapter 3, the First Department held that the ability to acknowledge legal duties or responsibilities is required for personhood. In response, the NhRP argued that this cannot be the right standard because it implies that humans who cannot acknowledge such duties or responsibilities, such as infants and comatose individuals, are not persons. The First Department addressed this objection by asserting that the excluded individuals "are still human beings, members of the human community," and that this is the basis of their personhood and legal protection (*Matter of Nonhuman Rights Project, Inc. v. Lavery* [2017]: 6). The goal of this chapter is to explore how this idea of "membership in the human community" constitutes a distinctive conception of personhood and to consider the implications for Kiko and Tommy.

One possible interpretation of "human community" puts the exclusive emphasis on 'human,' understood as a biological category, so that "human community" is a synonym for "members of the species *Homo sapiens*." This interpretation is indistinguishable from the species membership view addressed in Chapter 2, and so we set it aside as implausible. However, a second possible interpretation puts the emphasis on "community," referring to membership in the sorts of communities or societies that humans form. On this view, personhood cannot be achieved by an individual wholly on their own without any social contact or social context; rather, personhood is a relational condition that we achieve through development and recognition within a community of persons.

PERSONHOOD-BY-PROXY

There are two ways of understanding the idea of personhood as a relational status. One, which we can call the personhood-by-proxy view, invokes relational personhood as a kind of compensatory safety net to more traditional views of personhood that define it in terms of individual capacities. On this view, some individuals are persons by virtue of possessing the capacities deemed necessary for personhood (such as moral autonomy or legal responsibility). This is the direct or "front door" route to personhood, and those who qualify are deemed "self-standing persons" (Jaworska and Tannenbaum 2014, 2015) or "charter members" (Narveson 1987) of the personhood club. However, many of these persons have deep attachments of love and care to family members, friends, or neighbors who lack the relevant capacities to enter through the front door, even though they are recognized as integral members of the community. To ensure that the individuals to whom "self-standing persons" have special attachments are adequately protected, these "self-standing persons" extend personhood by association to all humans. For example, according to Narveson, humans who lack rational autonomy "are not, so to speak, charter members of the moral club," but in view of their "special attachments" to charter members, the latter will want to see them protected (Narveson 1987: 47). This, then, creates an indirect or "back door" route to personhood for those who lack the individual capacities deemed necessary for the direct or standard route to protected status. Personhood-by-proxy thus attempts to compensate for what would otherwise be the exclusion of any human who is not a self-standing person.

This view has been widely criticized by disability advocates and theorists, among others, for setting up a hierarchy of so-called real, normal, or "charter" persons, whose personhood is tied to their individual capacities and those who are given the protections of personhood "by courtesy or by proxy" (Wasserman et al. 2017). The view is objectionable because it makes the personhood of some of us dependent on our being the beneficiaries of the benevolence of those deemed *real* persons. History shows that "self-standing persons" have not always felt a strong attachment to those with disabilities, and, indeed, have often sought ways to disassociate from them or even to eliminate them through eugenics and sterilization (Silvers 2012). The status of persons-by-proxy is therefore precarious and vulnerable to shifting sentiments amongst their sponsors. It is also stigmatizing to implicitly or explicitly mark some individuals as deficient in relation to allegedly normative persons. For many philosophers and disability theorists, personhood is not a matter of courtesy or proxy. Any adequate account of personhood must acknowledge human diversity and affirm that each of us instantiates what it means to be a person on equal terms with everyone else. We do not have two tiers or levels. Every person is the moral equal of every other (Kittay 2005).

SOCIAL PERSONHOOD

The second way of understanding personhood as a relational status holds that community membership is not the back door to personhood but rather is itself the front door to personhood; it is how all of us become persons. We all come to be persons through embeddedness in interpersonal relationships of interdependency, meaning, and belonging. We can call this the 'social personhood' view. Social personhood is sometimes described as a "dyadic" view of personhood, as compared to a "monadic" conception that is based solely on the intrinsic capacities of individuals (Laitinen 2007).

This idea is captured in the Ubuntu philosophy of personhood stated as "I am because we are," in which personhood arises from participating in the social life of a community of persons, or, as stated in a traditional Zulu saying, "a person is a person through other people" (Eze 2010: 94). Persons do not exist as independent islands floating free of each other – and this is true for all of us. We are cooperative, interconnected beings who depend on the love, support, mutual recognition, purpose, and instruction we receive from each other (although the degree and forms of interdependency vary across individuals and over the course of our lifetimes). Individuals are persons by virtue of being embedded in webs of intersubjective and responsive relationships; our individual capacities emerge only because others nurture us as infants, teach us as toddlers, and cooperate with us as adults. And, here is the heart of the matter. This is true for all of us, equally, regardless of whether some of us appear as "self-standing" or not.

On this view, children, individuals with cognitive disabilities, or others who may be excluded from traditional conceptions of personhood, are indeed persons *in the same way* and *for the same reasons* as everyone else. On the first account of personhood, the fact that they may not be able to enter into contracts or bear certain legal responsibilities means they are not "self-standing persons," and, hence, are in need of proxy protections. But on a social personhood account, there are no "self-standing persons"—we are all persons through other people. All of us are dependent on others at some points in our lives and interdependent with others at all times. Regardless of our individual talents and capacities, if others fail to recognize and acknowledge us as equal persons, and to interact with us on this basis—due to biases based on race, ethnicity, class, gender, sexuality, ability, age, and so on—our personhood is undermined. Individuals with cognitive disabilities are no different in this respect. Neither are infants, toddlers, children, adolescents, or persons with mental illness or advanced dementia. They may lack some capacities of typical adults and the moral duties and citizenship responsibilities that accompany them. Nonetheless, all of them are fully embedded in the web of interpersonal relationships in which personhood is realized. They are, we can say, front door persons.

This social personhood view has been proposed, in different forms, by a wide range of ethical and religious traditions around the world. For example, various American Indigenous philosophies are characterized by relational ideas of personhood constructed through interagentivity (Grim 2006; Brighten 2011), as are some feminist and disability theories.[1] It may seem foreign to mainstream 'Western' legal jurisprudence, but in fact, it arguably sheds light on why arbitrary confinement is such a threat to persons and why habeas corpus is a crucial legal protection for ensuring our recognition as persons in community with others, safe from the arbitrary exercise of power to detain, confine, neglect, or isolate us from others. It also dovetails with recent attention to the devastating harms caused by solitary confinement and the fragmentation of personality caused by social isolation (Brownlee 2013; Guenther 2013). In other words, all persons need protection not only from arbitrary restrictions on their personal autonomy through detention but from any actions that restrict their ability to form and maintain the social relationships on which their personhood depends. Habeas corpus matters not just to protect an individual's autonomy and liberty but to also protect the ability to be a person through other people.

IMPLICATIONS FOR KIKO AND TOMMY

So far, we have distinguished two versions of the idea that personhood arises from membership in the human community: a person-by-proxy model which offers a backdoor to personhood for those members of society who cannot enter the front door by possessing certain cognitive capacities and a social personhood model that views membership in society as the front door by which everyone enters personhood. It is impossible to tell from the brief reference to "members of the human community" in the First Department's decision which of these models the court has in mind. However, we would argue that on either account, the court misinterprets the implications for Kiko and Tommy. The idea of community membership on either account actually *supports* including Kiko and Tommy in the community of persons.

Consider first the personhood-by-proxy account. We have mentioned critical weaknesses of this account. Nonetheless, even if the court accepts this view, it should grant Kiko and Tommy relief. To recall, this account states that "self-standing persons" (or the "charter members" of personhood), will extend personhood "by courtesy or by proxy" to other members of society whom they care about, whether as family members, friends, or neighbors. Defenders of this view —and perhaps the First Department—often assume that it is only human members of society who can benefit from this form of personhood-by-proxy. But there is nothing in the logic of personhood-by-proxy to ground this assumption. After all, "self-standing persons" often care deeply about nonhuman animals, and, indeed, view them as valued family members, friends, neighbors, or coworkers, and wish to include them in the community of persons. Indeed, the very existence of the NhRP and other efforts around the world to advance protection and recognition for chimpanzees are evidence that many humans wish to extend the protections of personhood to Kiko and Tommy and beings like them.

To be sure, there are also many humans who do not view nonhuman animals in this way. The harm currently being inflicted on Kiko and Tommy is evidence of this. But we must not forget that this has all too often been true, and continues to be true, in the human case as well. Demands to extend personhood-by-proxy beyond "charter members" are motivated precisely by the recognition that, without this protection, some individuals are vulnerable to harm and exploitation at the hands of others. The personhood-by-proxy account arises to protect those whose social recognition is uncertain or contested. The NhRP's suit on behalf of Kiko and Tommy is fully compatible with this logic, and illustrates how the personhood-by-proxy account supports Kiko's and Tommy's cases.

Consider now the social personhood account. Recall that on this view, personhood is the outcome of social relationships: we are persons through other persons. Defenders of this view often seem to assume that only humans can achieve social personhood. This implicitly rests on two assumptions: (1) that only human communities support the attribution of a social conception of personhood to their members, and (2) that only humans are members of "human communities."

Both assumptions should be questioned. The ethological evidence suggests that some free-living nonhuman animals form the sorts of communities that generate social personhood. Dolphins, for example, might well develop relations of social personhood within free-living dolphin communities, and chimpanzees might well become social persons within free-living chimpanzee societies even if there are no humans present. Ethological evidence gathered by primatologists suggests that free-living chimpanzees have a rich and complex social life which could be interpreted as bestowing personhood on its members. It is worth noting that many Indigenous views of social personhood take for granted that social personhood can arise within nonhuman animal communities and in inter-species relationships (see, e.g. Grim 2006; Brighten 2011).

However, we will not address this issue because it takes us away from the case of Kiko and Tommy who have been denied the opportunity to be born and raised within chimpanzee societies and have, instead, been brought into communities with humans. This brings us to the second assumption, namely, that only humans are members of human communities. This is an empirical, not a conceptual, question. Kiko and Tommy, like countless other animals, have been brought into human societies. They are not members of free-living chimpanzee communities but are embedded in interpersonal webs of dependency, meaning, and care with humans. Not only have we brought Kiko and Tommy into our human community, we have done so in ways that foreclose their option of being persons in a truly free-living chimpanzee community because they were not born and socialized into those communities. Thus, their social personhood will be nurtured (or suppressed) through their relations with humans. Like all persons, they become persons through their social relations with other persons, but in this case, these social relations include relations with both chimpanzees and humans. And as individuals whose social personhood is embedded in interpersonal relationships with humans and membership in a community with humans, they, too, must be protected when others exercise arbitrary power over their opportunities for social ties and the recognition and development of their social personhood.

A skeptic might ask: "Is it correct to say that we have brought Kiko and Tommy into human communities, given that they spend so much time in social isolation?" The answer must be yes. As already noted, humans have created the conditions which have denied to Kiko and Tommy the opportunity to be persons in community with free-living chimpanzees. Having done so, humans are obliged to ensure that Kiko and Tommy are recognized and able to develop as persons in the situations that we have created for them. While it is true that in practice we have failed miserably in our obligations to Kiko and Tommy as members of the human community, it would be perverse to use that failure as a basis to justify perpetuating this injustice.[2]

We have subjected Kiko and Tommy to the kind of arbitrary confinement and social deprivation that violates their personhood at a fundamental level. Recall, however, that this describes the historical situation for many human groups as well. History tells of too many humans who were incorporated into an oppressive community that detained, subjugated, deprived, and stigmatized them. Many of these humans sought habeas corpus relief. Indeed, this is one of the functions of habeas corpus: to protect members of the community who are being treated as things. This is the state in which we now find Kiko and Tommy.

In sum, while the social personhood conception emphasizes the importance of community, it does not follow that the community of persons is exclusive to human beings, not least because we have in fact brought other individuals, such as Kiko and Tommy, into our community. The question of which individuals are embedded in the social relationships that constitute our community is an empirical one, not deducible from mere biological classifications.

Insofar as one endorses a social personhood view, several implications follow for Kiko and Tommy. First, Kiko and Tommy are persons. Second, humans have denied Kiko and Tommy the opportunity to fully live and actualize as persons. Third, the relevant society in which Kiko and Tommy could live as persons is not an exclusively free-living chimpanzee society but must include humans. Fourth, Kiko and Tommy have the right to protection from arbitrary powers of detention, deprivation, or banishment which deny them the opportunity to fully develop and experience their personhood. And finally, the best remedy for Kiko and Tommy is to be released from captivity and into the rich social environment of an appropriate chimpanzee sanctuary, where their personhood can be more fully recognized and supported by humans and other chimpanzees.

LIMITS OF PERSONHOOD THROUGH COMMUNITY MEMBERSHIP

So far we have argued that on either the personhood-by-proxy or the social personhood account, Kiko and Tommy are clearly persons. Self-standing persons care about and wish to extend protection to Kiko and Tommy (as required by the personhood-by-proxy view), and their capacity to develop and flourish as individuals is tied up with their relations to human society (as required by the social personhood view).

While our focus has been on Kiko and Tommy, it seems clear that they are not the only nonhuman animals who are likely to qualify under community membership approaches. Both views are likely to include other animals captured or bred by humans in zoos or research facilities, as well as companion animals and farmed animals (e.g., dogs, cats, cows, and chickens) who have been brought into human society through the process of domestication.

One might ask whether there are any limits to whom or what can be included as persons on these views. For example, would either view attribute personhood to robotic toys simply because persons become attached to them? Consider the Aibo, a robotic dog first manufactured by Sony in 1999. Some owners claimed to form deep emotional bonds with these machines. When the robots began to break down and after Sony stopped making replacement parts, owners grieved for their lost pets. Indeed, a Buddhist temple in Japan conducted funeral services for a hundred such robots because their owners thought of their Aibos as ensouled beings (Inada 2017).

Would the Aibo qualify as persons on a community membership view? Perhaps they might on a personhood-by-proxy view in which individuals enter by the back door of personhood by virtue of being loved, cared for, and recognized by self-standing persons. This view puts the emphasis on the feelings and attitudes of so-deemed "normal" persons, and, therefore, is vulnerable to concerns that these attachments may be under- or over-inclusive in arbitrary ways. We noted earlier the concern that self-standing persons have often failed to feel attachment to people with cognitive disabilities and so failed to extend personhood-by-proxy to them.

The case of the Aibo suggests a different challenge: extending personhood-by-proxy to objects to which people are strongly attached. Some will regard the fact that robotic toys can be persons as a *reductio* of the personhood-by-proxy view.

On the social personhood view, however, the Aibo are not persons. The Aibo may be the object of human attachment, but they are not involved in intersubjective relationships with other persons. As we saw earlier, the purpose of recognizing personhood is to protect individuals from certain kinds of harms, including in particular the harm of social deprivation due to unlawful detention, deprivation, or banishment from community. It, therefore, only applies to the kinds of beings who can be harmed by the nature of their social engagement or social deprivation. Robots and other inanimate objects, like non-sentient life forms, cannot be persons on the social personhood view because they cannot be harmed by social deprivation.

While the social personhood view excludes inanimate objects and non-sentient life forms, it is still very wide in its conception of who qualifies as a member of the community of persons. Some philosophers would argue that it is too wide, and that some additional capacities are needed to qualify as a *bona fide* member of the community of persons. In other words, for someone to be a member of a community of persons requires more than social embeddedness and vulnerability to social exclusion. It requires that one possess certain additional individual traits or capacities that are more than sentience or vulnerability but less than the capacity to bear legal responsibilities.[3]

There are two possible kinds of traits these could be: biological or psychological. Biological traits are physical traits: having forty-six chromosomes, for example, or having a human father and mother. But this is merely a return to the view that only members of *Homo sapiens* qualify for personhood and, as argued in Chapter 2, restriction of personhood on the basis of species is arbitrary and unsupported by biological science. Psychological traits are mental capacities: having beliefs and desires, for example, or emotions, autonomy, and rationality. We will have more to say about these traits in Chapter 5, where we will discuss which psychological capacities may be necessary, and which sufficient, for personhood.

The goal of this chapter, however, is to distinguish the distinctive features of a community membership conception of personhood, one which does not merely collapse into either a species-based or capacity-based conception. The distinctive feature of the community membership conception is that it shifts the emphasis from individual traits and capacities to social relationships. The idea that personhood is a social achievement and that it is importantly linked to membership in the community is familiar and plausible. A community membership approach (especially the view which we refer to as social personhood) can indeed overcome some of the historic exclusions associated with other accounts of personhood, and so it is not surprising that the First Department may have appealed to it.

However, we need to think clearly and carefully about what that membership means and what it requires. We cannot simply assume that it excludes Kiko and Tommy. To be a person is to be embedded in social relationships of interdependency, meaning, and community. Kiko and Tommy clearly meet this criterion; we have made Kiko and Tommy part of our human community of persons by embedding them within relations of interdependence with us and, thus, rendering them vulnerable to forms of stigma, deprivation, or exclusion from this community. Kiko and Tommy are members of our community, and we owe them protection from the arbitrary power of others to define their social conditions.

NOTES

1 We should note, however, that some advocates of this approach avoid the language of personhood because of its historical tendency to prioritize or privilege individuals with so-called normal cognitive capacities. According to Quinn and Arstein-Kerslake (2012: 40), full inclusion of people with cognitive disability requires rejecting what they call "the 'myth-system' of personhood in human rights talk," and not everyone believes that it is possible to separate the term personhood from this "myth-system." Similar skepticism about the ability of 'personhood' to encompass the full diversity of human beings has been expressed by some post-colonial and critical race theorists (e.g., Dayan 2011). These theorists tend to use phrases like 'fellow beings' or 'fellow subjects' to emphasize the full diversity of humans who are owed social recognition of equal moral and legal standing.
2 Theorists of social personhood insist that while isolating individuals from the social community on which full development of their personhood depends can prevent them from becoming fully "actualized" persons, it in no way diminishes their moral standing as persons, and our obligations to them: "being recognized cannot be the precondition of the moral status or basic moral requirements, and it should rather be a response to those requirements" (Laitinen 2007: 260).
3 For an account of social personhood that insists on threshold capacities such as rational autonomy, see Laitinen (2007). On Laitinen's view, since these capacities can only be actualized through social embedding and recognition, personhood is dyadic not monadic. Most proponents of social personhood, however, see it as overcoming the exclusionary features of traditional personhood, recognizing that a much wider range of individuals can be persons in relations with other persons.

REFERENCES

Brighten, A. (2011). "Aboriginal peoples and the welfare of animal persons", *Indigenous Law Journal*, 10 (1): 39–72.

Brownlee, K. (2013). "A human right against social deprivation," *Philosophical Quarterly*, 63: 199–222.

Dayan, C. (2011). *The Law is a White Dog: How Legal Rituals make and unmake Persons*, Princeton: Princeton University Press.

Eze, M. O. (2010). *Intellectual History in Contemporary South Africa*, Basingstoke: Palgrave.

Grim, J. (2006). "Knowing and being known by animal: Indigenous perspectives on personhood" in P. Waldau and K. Patton (eds.) *A Communion of Subjects: Animals in Religion, Science, and Ethics*, New York: Columbia University Press, pp. 371–390.

Guenther, L. (2013). *Solitary Confinement: Social Death and Its Afterlives*, Minneapolis: Minnesota University Press.

Inada, H. (2017). "'Souls' of 100 Aibo robot dogs gain succor from Buddhist funeral" *Asahi Shimbun*, 9 June 2017. [Online] www.asahi.com/ajw/articles/AJ201706090040.html.

Jaworska, A. and Tannenbaum, J. (2014). "Person-rearing relationships as a key to higher moral status," *Ethics*, 124 (2): 242–271.

Jaworska, A. and Tannenbaum, J. (2015). "Who has the capacity to participate as a rearee in a person-rearing relationship?" *Ethics*, 125 (4): 1096–1113.

Kittay, E. F. (2005). "At the margins of moral personhood," *Ethics*, 116 (1): 100–131.

Laitinen, A. (2007). "Sorting out aspects of personhood: Capacities, normativity and recognition", *Journal of Consciousness Studies*, 14 (5): 248–270.

Matter of Nonhuman Rights Project, Inc. v. Lavery [2017] NY Slip Op 04574.

Narveson, J. (1987). "On a case for animal rights," *The Monist*, 70 (1): 31–49.

Quinn, G. Arstein-Kerslake, A. (2012). "Restoring the 'Human' in 'Human Rights': Personhood and Doctrinal Innovation in the UN Disability Convention", in C. Gearty and C. Douzinas (eds.) *Cambridge Companion to Human Rights Law*, Cambridge: Cambridge University Press.

Silvers, A. (2012). "Moral status: what a bad idea!" *Journal of Intellectual Disability Research*, 56 (11): 1014–1025.

Wasserman, D., A. Asch, J. Blustein, and D. Putnam (2017). "Cognitive disability and moral status," *The Stanford Encyclopedia of Philosophy* (Fall 2017 Edition). [Online] https://plato.stanford.edu/archives/fall2017/entries/cognitive-disability/.

The capacities conception
Five

The NhRP argues that the capacity for autonomy is *sufficient*, but not necessary, for personhood. The court rulings don't dispute this claim, nor do they challenge the empirical claim that chimpanzees are autonomous. In this chapter, we defend the NhRP's claim about autonomy. Our discussion will only consider how autonomy and personhood are discussed in the 'Western' philosophical tradition. We reject conceptions of personhood that exclude those humans we take to be uncontroversially persons (understood to cover humans from birth to death) and those that favor a hierarchy of personhood, as they do not protect the interests of all those humans as moral equals.

In the first part of the chapter we present the commonly cited capacities relevant to the possession of personhood and show that the concept of personhood that emerges can be interpreted as either an "essentialist concept" or a "cluster concept." Since the essentialist version is under-inclusive (it excludes some humans as persons) while the cluster version does not, we reject the former and endorse the latter. The cluster version explains why the capacity for autonomy is sufficient for personhood. With a proper cluster conception of persons, it is evident that there are different ways of being a person. Different humans can be persons in different ways. Members of different species can be persons, too. In the second part of this chapter, we argue that chimpanzees are autonomous when autonomy is understood more inclusively. We conclude by considering what sorts of protections a chimpanzee should have given the kind of persons they are.

CAPACITIES OF PERSONHOOD

In 'Western' philosophy, there is a long history of thinking about personhood in terms of capacities, going back at least to Locke. He described what it is to be a person this way: "a thinking intelligent being that has reason and reflection and can consider itself as itself, the same thinking thing in different times and places; which it does only by that consciousness which is inseparable from thinking and…essential to it" (Locke 1689: 302).

Contemporary philosophical discussions of personhood offer expanded lists of capacities relevant to personhood. The capacities often cited include:

Sentience, often associated with basic awareness[1]
Emotions, including happiness, empathy, sadness, fear, anger, or pain[2]
Autonomy, the ability to act on behalf of oneself, including exercising executive control over the formation of one's goals and the means for achieving them[3]
Self-awareness, of one's own mental life[4]
Sociality, in relation to other individuals[5]
Language, used to communicate to others and self[6]
Rationality, means-end reasoning or logical thought processes[7]
Narrative self-constitution, thinking of oneself as a persisting subject with past experiences, a character in one's own story who will author one's future experiences[8]

> Morality, an understanding of what is good, right, or virtuous[9]
> Meaning-making, a vision of a life worth pursuing or a sense of what it is to live well[10]

While some of the properties on this list are disputed, and as a group we could not endorse all of them, this list reflects common ways in which 'Western' philosophy has tended to conceptualize persons. Regardless of the properties one includes, there are two ways to look at capacities like these:[11] as *essential* features of persons or as *clusters* of properties that are variously constitutive of persons.

On essentialist versions, a person satisfies a set of individually necessary and jointly sufficient conditions. Individuals must possess all in the set of conditions to count as persons. Essentialism thus denies that some humans are persons, since some humans never possess some of the traits or fail to have some of them at some point in their lives. Since we are working within a framework that is friendly to legal views that all humans from birth to death are persons, we reject the essentialist version.

On cluster conceptions, individuals must have *some*—and *tend* to have *several*—of the personhood traits, but no one of the traits is required. A variety of subsets of capacities are regarded as sufficient but not necessary for personhood. Cluster kinds consist of "a complicated network of similarities overlapping and criss-crossing" (Wittgenstein 1967: 66), and they are widely found in the social world. We endorse this approach because, depending on the list of capacities, it need not exclude any humans as persons and permits different personhood profiles for different individuals. It also ensures that all human beings as we understand them remain persons and are equally regarded as such. Importantly, it also does not introduce *ad hoc* exclusions of other beings who meet the criteria.

PERSONHOOD AND AUTONOMY

The NhRP's case is based on one particular capacity—autonomy—and for good reason. The philosophical conception of personhood is often framed in terms of autonomy, understood as the ability to act on behalf of oneself, including exercising executive control over the formation of one's goals and the means for achieving them. Its importance in contemporary ethics can be traced to Kant's conception of persons as autonomous agents who are ends in themselves. Philosopher John Christman echoes this view when he writes, "to be autonomous is to be one's own person, to be directed by considerations, desires, conditions, and characteristics that are not simply imposed externally upon one, but are part of what can somehow be considered one's authentic self" (Christman 2015). Since to be autonomous is to be one's own person, to be autonomous is sufficient to be a person on such views.

In political theory, autonomy is associated with a number of benefits (Christman 2015). For example, autonomy is sometimes taken to be sufficient for political standing. For many, having autonomy also means that one should not be treated paternalistically. Autonomy is also associated with a number of properties in moral theory. Autonomous agents are sometimes seen as morally responsible for their actions, and autonomous agents are taken to be intrinsically valuable, ends in themselves (Kant [1785] 2001).

These benefits don't always come with autonomy as defined above because there are more and less demanding views of it. On a very demanding view, autonomy as normative self-governance requires the metacognitive ability to abstractly consider principles of action and judge them as acceptable or unacceptable (Johnson and Cureton 2018). Philosopher

Christine Korsgaard argues that other animals lack autonomy in this sense because normative self-governance requires a certain form of self-consciousness:

> ... namely, consciousness of the grounds on which you propose to act *as grounds*. What I mean is this: a nonhuman agent may be conscious of the object of his fear or desire, and conscious of it as *fearful* or *desirable*, and so as something to be avoided or to be sought. This is the ground of his action. But a rational animal is, in addition, conscious *that* she fears or desires the object, and *that* she is inclined to act in a certain way as a result. That's what I mean by being conscious of the ground *as a ground*. She does not just think about the object that she fears or even about its fearfulness but about her fears and desires themselves.
>
> (Korsgaard 2006: 113)

For Korsgaard, autonomy as normative self-governance requires the ability to realize *that* one fears or desires some state of affairs, which, in turn, requires metacognitive capacities such as the ability to accurately represent the contents of one's own beliefs and desires and the ability to evaluate those commitments. This more demanding version of autonomy is closely related to having moral or legal *responsibility*. On this view, any humans who do not have the capacity to engage in normative self-governance will not count as autonomous. This account of autonomy fails to identify the autonomous actions of most typical human adults, unlike the more general way of understanding autonomy as the ability to act on behalf of oneself. Research on adult moral reasoning suggests that adults do not generally consider their reasons when making moral judgments (Haidt 2001), although their accompanying actions are still commonly regarded as autonomous. Research on adult action suggests that we can fail to know our own reasons for our seemingly autonomous actions or judgments. Social psychologists, Richard Nisbett and Lee Ross, found that adults will confabulate their reasons for action, and do not have direct access to some of their action-guiding processes (Ross and Nisbett 1991). On Korsgaard's more demanding account, many humans, and certainly children, would lack autonomy, and many of the actions we think are freely chosen would not be autonomous.

Consider a parallel case of more and less demanding accounts of having language in which the more demanding account involved producing poetry. While producing poetry is at the pinnacle of language capacities and is sufficient evidence for having language, it is not necessary. Likewise, we find the demanding account of autonomy describes the pinnacle of self-directed action. For these reasons, the more demanding accounts of autonomy are less helpful for constructing a capacities account of personhood.

On a weaker understanding of autonomy tied to moral or legal *standing*, rather than moral or legal responsibility, children's less developed self-governance can be consistent with being autonomous in the sense that they are the source of their own actions, and they act on behalf of themselves rather than because of some external force or internal compulsion. From an early age, humans form preferences, set goals, and act so as to satisfy wants. They are agents who can act voluntarily. They are cognitively flexible beings who can form their own goals and act rationally to achieve those goals. There are clear contrasts in some other animals. The harvester ant, for example, isn't acting autonomously when she takes dead ants to the refuse pile; rather she is automatically responding to the oleic acid that ant corpses produce as they decompose. When E.O. Wilson treated live ants with oleic acid, he observed that they were carried "alive and kicking" to the refuse pile by their sisters. The movement of the coated ants did not override the behavioral response of their sisters to the stimulus of oleic acid (Wilson 1985). The conceptual space between normative self-governance and automatic, controlled behavior is not visible in the more demanding accounts of autonomy.

A distinct advantage of a less demanding view of autonomy is that it does not violate the common understanding that most ordinary human behavior is autonomous. Bioethicist and philosopher Tom Beauchamp and comparative psychologist Victoria Wobber offer one such account. They propose that an act is autonomous if an individual self-initiates an "action that is (1) intentional, (2) adequately informed…and (3) free of controlling influences" (Beauchamp and Wobber 2014: 119). Conceptions of autonomy like this support chimpanzee autonomous capacity.

AUTONOMY IS SUFFICIENT FOR PERSONHOOD ON A CLUSTER CONCEPT OF PERSONS

A second reason to focus on autonomy is that as a concept it implies various capacities that are identified with inclusive analyses of personhood. As highlighted by Beauchamp and Wobber, to be autonomous is to have two core capacities: the capacity to act intentionally (which assumes capacities to form goals and direct one's behavior accordingly) and to be adequately informed (which assumes capacities to learn, to make inferences, and acquire knowledge through rational processes) (Beauchamp and Wobber 2014). Many of the capacities listed above are also implicated in these two core capacities. The capacity to experience emotions is used to form goals. Sentience is typically associated with both responsive behavior and gaining information about the world that can be used to order future behavior. Rational processes can be entangled with setting goals and working toward them, as well as with acquiring information about one's physical and social worlds, and revising such information should its inaccuracy become apparent. Self-awareness helps one realize what one needs to know in order to achieve a goal and what one does not already know. Since an autonomous capacity implies other person-making capacities, evidence of autonomy is sufficient evidence of personhood considered as a cluster concept. Thus, an individual can be identified as a person on autonomy grounds alone. However, one can still be a person *without* being autonomous.

SCIENTIFIC EVIDENCE FOR CHIMPANZEE PERSONHOOD CAPACITIES

Since the first chimpanzee field research sites were started by Jane Goodall in Gombe, Tanzania in 1960 and Toshisada Nishida in Tanzania's Mahale Mountains in 1965, we have learned much about chimpanzees' cognitive and social capacities. Just as there are individual differences between humans and between various groups of humans, there are individual differences between chimpanzees (King and Figueredo 1997) and between various groups of chimpanzees (Whiten et al. 1999). These inter-group differences permit identification of individuals as members of particular groups (for researchers and perhaps for local chimpanzees) and many primatologists speak of distinct group cultures. We will canvass some of what is known about chimpanzees that connects with capacities listed above.

Autonomy

Chimpanzees are cognitively flexible animals whose actions are not merely fixed responses to situations. They can act intentionally, and they can plan and act so as to achieve goals. For example, chimpanzees have been observed planning attacks on annoying zoo visitors (Osvath 2009; Osvath and Karvonen 2012) and planning where they will eat breakfast (Janmaat et al. 2014). In formal experiments chimpanzees demonstrate that they engage in means-end reasoning about their goals. As an example, they are able to delay gratification as indicated by their performance in a chimpanzee version of the marshmallow task. Chimpanzees choose to pass on an immediate tasty snack in order to receive a larger one later; they will even self-distract in order to facilitate self-control (Beran et al. 2014). Chimpanzees can use what they know and don't know to shape their choices. For example, chimpanzees can use probabilities to make decisions about which container to choose food from, given the relative distributions of preferred food items (Rakoczy et al. 2014). When chimpanzees don't have enough information to solve a task, they seek the information needed (Beran et al. 2013; Bohn et al. 2017). And when they know that they are right, they anticipate getting a reward and position themselves to quickly receive it (Beran et al. 2015).

Emotions

Chimpanzees are emotional beings, expressing such emotions as excitement, anger, empathy and fear. Scientists look at animal behavior, physiology, and social interaction in their studies of chimpanzee emotions (de Waal 2011). The developmental comparative psychologist Kim Bard has documented chimpanzee emotional development, and finds that chimpanzee infants, like human infants, develop their emotional responses in social contexts, for example smiling in response to a caregiver's smile or to a caregiver tickling and vocalizing (Bard 2012).

Language

While chimpanzees cannot master human languages and do not have a chimpanzee language with the same kind of grammatical structure as human languages, they do have communication systems and they can learn to use some elements of human symbolic communicative systems. Washoe, one of a number of chimpanzees used in experiments that exposed them to learning environments typical of human children, was taught some American Sign Language and used ASL signs to communicate to humans and other chimpanzees for the rest of her life (Fouts 1998). Other chimpanzees have been successfully taught a lexicon symbolic communication system (Savage-Rumbaugh et al. 1998).

Chimpanzees have their own natural communicative system consisting of vocalizations and gestures. From observations of wild chimpanzees, we know they have alarm calls, food calls, and calls that express emotions (Goodall 1986). The gestural system of chimpanzees appears to be shared with other great apes. For example, when a chimpanzee or a bonobo shows a partner an exaggerated self-directed scratch, the partner initiates grooming the signaler (Graham et al. 2018). Evidence that chimpanzee communication is intentional comes from finding that they will repeat and elaborate on a message when initial attempts to deliver the message fails (Leavens et al. 2005).

Sentience

Chimpanzees have the same neuroanatomical structures and mechanisms as other mammals, including various sensory systems, such as those implicated in the sensory and affective components of pain (Knight 2008). As noted by de Waal, chimpanzees essentially have smaller versions of human brains. No new structure in human brains is lacking in chimpanzee brains (de Waal 2016). Scientists widely accept that animals such as chimpanzees are conscious experiencers of their lives as evidenced by the signing of The Cambridge Declaration on Consciousness in 2012.[12]

Rationality

Tool creation and use indicates the existence of means-end, or instrumental, rationality. Various chimpanzees make and use tools including stones, anvils, and wedges for nut cracking, puncture tools for opening termite mounds, fishing tools for acquiring termites and ants, leaves as sponges to collect water, and more (Shumaker et al. 2011). Reasoning involves the ability to understand what follows from the information one already has. Experiments demonstrate that chimpanzees and other great apes can understand abstract relationships between material objects involving the notions of "same" and "different" (Tomasello and Call 1997). They solve tasks that suggest they may be engaged in reasoning by exclusion, finding treats hidden under cups in a kind of a shell game (Call 2004).

Self-awareness

The psychologist Gordon Gallup introduced the mirror self-recognition test for self-awareness, surreptitiously marking chimpanzee and human infants and exposing them to a mirror. When a marked subject touches the mark more frequently when there is a mirror available than when there is not, the subject passes the test and so demonstrates self-awareness (Gallup 1970). Chimpanzees are among the species who can pass this test, though of course, not all individual chimpanzees in any given test pass.

Sociality

Chimpanzees are social beings, and have robust social cognitive capacities. They take other chimpanzees to be agents, recognizing intentions in others. Chimpanzees can distinguish between unwilling and unable actors (Call et al. 2004) and between helpers and hinderers (Krupenye and Hare 2018). Chimpanzees pass the false belief task, long thought to be evidence of having a theory of mind (Krupenye et al. 2016; Buttelmann et al. 2017); they can predict others' behavior and anticipate others' desires even when others have a false belief. Chimpanzees also show sensitivity to others' informational states and will inform ignorant chimpanzees, but not knowledgeable chimpanzees, about the presence of fearful stimuli like snakes (Crockford et al. 2012). Chimpanzees are able to navigate quite complex physical and social worlds, acting intentionally and seeking information as needed. Coalitions and alliances among a subset of chimpanzees in a group, which are a widespread social strategy in chimpanzee communities, evince capacities to coordinate behavior, realize preferences (including preferences about treatment), and secure social influence (Mitani et al. 2010). There is evidence that among free-living chimpanzees food sharing is based on a score-keeping form of reciprocity (Jaeggi and Gurven 2013) and that chimpanzees hunt monkeys cooperatively (Boesch 1994).

Not all scientists agree about all capacities of chimpanzees. The comparative developmental psychologist Michael Tomasello, for example, argues that chimpanzees lack reciprocity, joint action, and cooperation based on results of his cognitive studies on *captive* chimpanzees (Tomasello 2016). We expect that further research on behaviors related to reciprocity will lead to better understanding of the extent to which chimpanzees are like and unlike human groups where cooperation and exchange are concerned.

AUTONOMY, WELFARE, AND SANCTUARIES

That chimpanzees are autonomous in the ways we have described has clear connections with three considerations that bring together animal welfare, or well-being, and ethics. First, "choice" and "control" are increasingly viewed as important in designing captive environments to better meet the needs of cognitively and emotionally complex nonhuman animals (Maple and Perdue 2013). Second, adequate sanctuary environments can have a rehabilitative effect on chimpanzees who express behavioral problems due to inadequacies in their past captive environments and harms[13] they have experienced (Wobber and Hare 2011). Third, autonomy foregrounds harms that arise when an individual is coerced, deceived, forced, or manipulated into actions that they would not otherwise choose given their preferences and values (Beauchamp and Wobber 2014). This section will discuss the first two of these considerations while the next section will discuss the third.

That chimpanzees make choices and have some control over their daily lives reflects the kinds of capacities chimpanzees enjoy. As already indicated, free-living chimpanzees live in complex social worlds. Their communities can range from a handful of members (Stumpf 2011: 347) to well over 160 (Stanford 2018). The fission-fusion style of social life seen in free-living chimpanzees permits a great deal of choice and control, including getting distance from particular members of the community, choice in sexual encounters (perhaps outside the watchful eye of higher ranking chimpanzees within the community or with chimpanzees from rival communities), choice of where to forage for food, or whether to explore or play rather than feed (Stumpf 2011). The rationality and autonomy of chimpanzees are utilized to make the choices that they do.

Choice and control in chimpanzee social lives also impacts their emotional well-being. The social lives of free-living chimpanzees are not dissimilar from our own. They can form strong and stable friendship bonds with community members and have preferences about the company they keep and who they will groom. Because chimpanzee males do not typically leave their natal community, the bonds between males can be particularly strong, and family bonds between siblings as well as children and mothers play a significant social role. Primatologists have uncovered the social role of coalitions and alliances that provide their members access to food, receptive members of their community, or even positions of social influence (Stumpf 2011). Conflicts can be reasonably interpreted as social moments where preferences are expressed, and resolution or consolation as social moments where comfort is provided and valued relationships are affirmed (de Waal 1996). Through social learning, chimpanzees come to understand how to behave around others with whom they are living, who enjoys what friendships and allegiances, and who has greater social influence or dominance (Matsuzawa 2010; Mitani et al 2010). This information is used to choose with whom to affiliate, which in turn impacts chimpanzees' emotional regulation and stress hormones (Wittig et al. 2016).

The daily lives of free-living chimpanzees are filled with choices, both great and small, and choices carry greater physical and social significance as chimpanzees age in their communities. For example, the nature of dependency changes as juveniles mature and become both more independent and responsible for their actions (e.g., adult tolerance for how young chimpanzees behave changes as they age) and females typically leave their communities around puberty to find new ones (Humle 2006; Stumpf 2011). Chimpanzees not only exercise choice; they must learn not to behave impulsively. Responding aggressively to a higher ranked chimpanzee or group of chimpanzees can generate conflict and result in personal injury or, in the extreme, isolation. Seeking a sexual encounter with a receptive member of the community can lead to conflict if care is not taken to avoid the watchful eye of a higher ranked male chimpanzee (de Waal 1998). Vocalizations that alert members of the community to sources of food, if unmodulated, can deny a chimpanzee first access to food or the ability to eat to their satisfaction (e.g., Brosnan and de Waal 2002). These all bear a relation to self-governance which is a hallmark of autonomy as we have described it.

The cognitive and affective complexity of chimpanzee life is tied to welfare needs for greater choice and control within captive environments. In zoos and biomedical laboratories, it is increasingly recognized that boredom can have profound and detrimental effects on an animal's physical and psychological well-being. Stereotypies, which are abnormal behaviors such as body rocking, repetitive behaviors, and self-injury, are recognized as indicators of poor welfare and are thought to be ways for such animals to cope with impoverished (and sometimes stressful) environments (Maple and Perdue 2013). Forcing animals to comply with demands, instead of training them to cooperate by reinforcing the desired behaviour (such as moving cages or presenting body parts for veterinary care) causes significant stress that can later be expressed in aggression, stereotypies, or social isolation (Bloomsmith et al. 2007; Perlman et al. 2010, 2012; Wolfensohn and Lloyd 2013). Some primatologists have also discovered that chimpanzees enjoy and seek out puzzles and tasks that they can solve or successfully accomplish. Providing opportunities to choose—such as choosing where they hang out and with whom, when to eat a snack or sleep—helps relieve boredom and reduce stress (Ross 2010).

Adequate sanctuaries provide such opportunities for choice and control and provide environments in which chimpanzees who have come from places where they have experienced significant harms and social isolation can begin to heal. They can live with other chimpanzees in ways that over time permit them to develop skills, capacities, and knowledge that better reflect what it means for a chimpanzee to be a chimpanzee. To illustrate, consider one of the largest sanctuaries in the United States, Chimp Haven, which resides on 200 acres, and is currently home to over 200 chimpanzees.[14] Chimpanzees are housed socially which means they get to socially interact in group sizes currently ranging from 9 to 23. They are provided climbing structures, toys, and foraging puzzles (e.g., artificial 'termite mounds' containing treats), some of which provide them with tool-learning opportunities. Because they have some large enclosures (such as three- to five-acre forested habitats), individuals can distance themselves from other chimpanzees and humans. The habitats also provide opportunities to climb, explore, and relax. Their veterinarians use positive reinforcement training to secure the cooperation of the resident chimpanzees. This minimizes the use of sedation which, for geriatric chimpanzees, can be dangerous and is also far less stressful to the chimpanzees.[15]

AUTONOMY AND ETHICS

Autonomous individuals have a basic interest in exercising their autonomy (Beauchamp and Childress 2001). Violating a basic interest is widely regarded as a harm, so violating someone's autonomy is a harm.

It is important to contrast the development and exercise of autonomous capacities of free-living chimpanzees with what is possible for chimpanzees in many captive environments. Zoos can impose significant restrictions on movement, the size of the community, the daily opportunities for choice, and with whom chimpanzees interact. Biomedical laboratories do the same, but the nature of the work and the need to control factors that can confound scientific results has meant that lab chimpanzees often live in even more socially and physically impoverished environments. Such chimpanzees face stressors associated, not just with standard veterinary care or the presence of humans, but with various protocols shaped by the relevant studies in which they are used (for some examples see Bloomsmith and Else 2005). Although the life prospects for biomedical laboratory chimpanzees in the United States are changing, there are still many currently housed in the laboratories where they were used in harmful research.

The Fourth Department ruled, and the First Department concurred, that habeas corpus relief is not available to Kiko or Tommy because the NhRP is not seeking their release from captivity but rather their relocation to a suitable sanctuary. This judgment heavily depends on lumping together markedly different kinds of captivity. Our discussion of autonomy provides a way to usefully compare Kiko's and Tommy's current captive conditions to those they could enjoy in a sanctuary. Both Kiko and Tommy are currently housed alone and in small enclosures. Should they be relocated to sanctuaries such as Chimp Haven or Save the Chimps[16], several things would change. They would no longer be housed alone and would no longer be confined indoors. They would have markedly more freedom to roam, explore, and forage. They would have the opportunity to develop and exercise more typical chimpanzee social capacities and expand their goals and preferences to reflect the greater opportunities afforded them. All of these changes would allow Kiko and Tommy the kind of choice and control they currently lack, which would satisfy their basic interests. In their current conditions of captivity, their interests in acting autonomously are profoundly violated. A sanctuary such as Chimp Haven or Save the Chimps promises not only much greater freedom but also a setting where their autonomous capacities would be respected.

CONCLUSION

Kiko and Tommy are persons because they are autonomous. This is the philosophical essence of the NhRP's legal position and the reason that the NhRP holds that these chimpanzees are entitled to protection as persons. On the most plausible philosophical account of autonomy and the most accurate reading of the scientific literature, the NhRP's argument is sound. Our argument for this employs a cluster conception of personhood widely accepted among philosophers as superior to essentialist conceptions because it can recognize the personhood and protect the equal moral status of all human persons. Once we appreciate the virtues of the cluster conception of personhood, we cannot deny that individuals of species other than *Homo sapiens* may be persons according to that conception. Kiko and Tommy are paradigmatic examples of such individuals. We owe them the kind of consideration all persons deserve.

NOTES

1. For example, Chan and Harris (2011); Rowlands (2016); Dennett (1976); Locke (1690); Strawson (1959); Varner (2012); and Warren (1973).
2. For example, Dennett (1976) and Warren (2014).
3. For example, Dennett (1976); Frankfurt (1971); Locke (1690); and Rawls (1971).
4. For example, Chan and Harris (2011); Dennett (1976); Locke (1690); Varner (2012); Rowlands (2016); and Warren (2014).
5. For example, Chan and Harris (2011); Dennett (1976).
6. For example, Chan and Harris (2011); Dennett (1976); Varner (2012); and Warren (2014).
7. For example, Chan and Harris (2011); Dennett (1976); Locke (1690); Varner (2012); and Warren (2014).
8. For example, Schechtman (1996); Varner (2012); and Comstock (2017).
9. For example, Dennett (1976); Darwall (2006); and Warren (2014).
10. For example, Chan and Harris (2011); Frankfurt (1971); Rawls (1971); and Wolf (2010).
11. It is important to note that for many of these capacities there is no shared understanding of what the capacity amounts to, and the authors cited do not always agree with one another about what the capacity term refers to. For example, Rowlands (2016) explicitly distinguishes his view of self-awareness from that of Locke's and others. This will be apparent in the discussion of autonomy in the next section.
12. See fcmconference.org/img/CambridgeDeclarationOnConsciousness.pdf.
13. For our purposes here, we follow Beauchamp in describing a harm as a setting back of the interests of an individual that they otherwise would

not experience, as a result of another's acts of commission or omission, or a naturally occurring event (Beauchamp 2011, 2014).
14 See www.chimphaven.org/.
15 These details are found on their online videos: https://chimphaven.org/videos/.
16 Save the Chimps is currently home to 248 chimpanzees residing on twelve three-acre islands. See www.savethechimps.org/.

REFERENCES

Bard, K.A. (2012). "Emotional engagement: How chimpanzee minds develop," in F.B.M. de Waal and P.F. Ferrari (eds.) *The Primate Mind: Built to Connect with Other Minds*, Cambridge, MA: Harvard University Press, pp. 224–245.

Beauchamp, T.L. and Wobber, V. (2014). "Autonomy in chimpanzees," *Theoretical Medicine and Bioethics*, 35 (2): 117–132.

Beran, M.J., Smith, J.D., and Perdue, B.M. (2013). "Language-trained chimpanzees (Pan troglodytes) name what they have seen but look first at what they have not seen," *Psychological Science*, 24 (5): 660–666.

Beran, M., Evans, T.A., Paglieri, F., McIntyre, J.M., Addessi, E., and Hopkins, W.D. (2014). "Chimpanzees (Pan troglodytes) can wait, when they choose to: A study with the hybrid delay task," *Animal Cognition*, 17 (2): 197–205.

Beran, M.J., Perdue, B.M., Futch, S.E., Smith, J.D., Evans, T.A., and Parrish, A.E. (2015). "Go when you know: Chimpanzees' confidence movements reflect their responses in a computerized memory task," *Cognition*, 142: 236–246.

Bloomsmith, M.A. and Else, J.G. (2005). "Behavioral management of chimpanzees in biomedical research facilities: The state of the science," *ILAR Journal*, 46 (2): 192–201.

Bloomsmith, M.A., Marr, M.J., and Maple, T.L. (2007). "Addressing nonhuman primate behavioral problems through the application of operant conditioning: Is the human treatment approach a useful model?" *Conservation, Enrichment and Animal Behaviour*, 102 (3): 205–222.

Boesch, C. (1994). "Cooperative hunting in wild chimpanzees," *Animal Behaviour*, 48 (3): 653–667.

Bohn, M., Allritz, M., Call, J., and Völter, C.J. (2017). "Information seeking about tool properties in great apes," *Scientific Reports*, 7 (1): 10923.

Brosnan, S.F. and de Waal, F.B.M. (2002). "Regulation of vocal output by chimpanzees finding food in the presence or absence of an audience," *Evolution of Communication*, 4 (2): 211–224.

Buttelmann, D., Buttelmann, F., Carpenter, M., Call, J., and Tomasello, M. (2017). "Great apes distinguish true from false beliefs in an interactive helping task" PLOS ONE, 12(4): e0173793.

Call, J. (2004). "Inferences about the location of food in the great apes (Pan paniscus, Pan troglodytes, Gorilla, and Pongo pygmaeus)," *Journal of Comparative Psychology*, 118 (2): 232–241.

Call, J., Hare, B., Carpenter, M., and Tomasello, M. (2004). "'Unwilling' versus 'unable': chimpanzees' understanding of human intentional action," *Developmental Science*, 7 (4): 488–498.

Chan, S. and Harris, J. (2011). "Human animals and nonhuman persons," in T.L. Beauchamp and R.G. Frey (eds.) *The Oxford Handbook of Animal Ethics*, New York: Oxford University Press, pp. 304–331.

Christman, J. (2015). "Autonomy in moral and political philosophy," in E. N. Zalta (ed.) *Stanford Encyclopedia of Philosophy* (Spring 2018 edition). [Online] https://plato.stanford.edu/archives/spr2018/entries/autonomy-moral/ [Accessed 4 April 2018].

Comstock, G. (2017). "Far-persons," in A. Woodhall and G. Garmendia da Trindade (eds.) *Ethical and Political Approaches to Nonhuman Animal Issues*, New York: Palgrave Macmillan, pp. 39–71.

Crockford, C., Wittig, R.M, Mundry. R., and Zuberbühler, K. (2012). "Wild chimpanzees inform ignorant group members of danger," *Current Biology*, 22 (2): 142–146.

Darwall, S. (2006). *The Second-Person Standpoint: Morality, Respect, and Accountability*, Cambridge, MA: Harvard University Press.

de Waal, F.B.M. (1996). "Conflict as negotiation," in W.C. McGrew, L.F. Marchant, and T. Nishida, (eds.) *Great Ape Societies*, Cambridge, MA: Cambridge University Press, pp. 159–172.

de Waal, F.B.M. (1998). *Chimpanzee Politics: Power and Sex among Apes*. Baltimore, MD: The Johns Hopkins University Press.

de Waal, F.B.M. (2011). "What is an animal emotion?" *Annals of the New York Academy of Sciences*, 1224 (1): 191–206.

de Waal, F.B.M. (2016). *Are We Smart Enough to Know How Smart Animals Are?* New York: W. W. Norton & Company.

Dennett, D.C. (1976). "Conditions of personhood," in A. Rorty (ed.) *The Identities of Persons*, Berkeley, CA: University of California Press, pp. 175–196.

Fouts, R. (1998). *Next of Kin: My Conversations with Chimpanzees*, New York: William Morrow Paperbacks.

Frankfurt, H.G. (1971). "Freedom of the will and the concept of a person," *The Journal of Philosophy*, 68 (1): 5–20.

Gallup, G. (1970). "Chimpanzees: Self-recognition," *Science*, 167: 341–343.

Goodall, J. (1986). *The Chimpanzees of Gombe: Patterns of Behavior*. Cambridge, MA: Harvard University Press.

Graham, K.E., Hobaiter, C., Ounsley, J., Furuichi, T., and Byrne, R.W. (2018). "Bonobo and chimpanzee gestures overlap extensively in meaning," *PLoS Biology*, 16 (2): 1–18.

Haidt, J. (2001). "The emotional dog and its rational tail: A social intuitionist approach to moral judgment," *Psychological Review*, 108: 814–834.

Humle, T. (2006). "Ant dipping in chimpanzees: An example of how microecological variables, tool use, and culture reflect the cognitive abilities of chimpanzees," in T. Matsuzawa, M. Tomonaga, and M. Tanaka (eds.) *Cognitive Development in Chimpanzees*, Japan: Springer, pp. 452–475.

Jaeggi, A.V. and Gurven, M. (2013). "Reciprocity explains food sharing in humans and other primates independent of kin selection and tolerated scrounging: a phylogenetic meta-analysis," *Proceedings: Biological Sciences*, 280 (1768): 1–8.

Janmaat, K.R.L., Polansky, L., Ban, S.D., and Boesch, C. (2014). "Wild chimpanzees plan their breakfast time, type, and location," *Proceedings of the National Academy of Sciences of the United States of America*, 111 (46): 16343–16348.

Johnson, R. and Cureton, A. (2018) "Kant's moral philosophy,", *The Stanford Encyclopedia of Philosophy*, Edward N. Zalta (ed.), URL = <https://plato.stanford.edu/archives/spr2018/entries/kant-moral/>.

Kant, I. (1785). "The groundwork of the metaphysics of morals," in M.J. Gregor (ed.), *Practical Philosophy, The Cambridge Edition of the Works of Immanuel Kant*, Cambridge University Press, Cambridge, pp. 37–108.

King, J.E. and Figueredo, A.J. (1997). "The five-factor model plus dominance in chimpanzee personality," *Journal of Research in Personality*, 31 (2): 257–271.

Knight, A. (2008). "The beginning of the end for chimpanzee experiments?" *Philosophy, Ethics, and Humanities in Medicine: PEHM*, 3: 16.

Korsgaard, C.M. (2006). "Morality and the distinctiveness of human action," in S. Macedo and J. Ober (eds.) *Primates and Philosophers: How Morality Evolved*, Princeton, NJ: Princeton University Press, pp. 98–119.

Krupenye, C., Kano, F., Hirata, S., Call, J., and Tomasello, M. (2016). "Great apes anticipate that other individuals will act according to false beliefs," *Science*, 354(6308): 110–114.

Krupenye, C. and Hare, B. (2018). "Bonobos prefer individuals that hinder others over those that help," *Current Biology*, 28 (2): 280–86.e5.

Leavens, D., Russell, J. L., and Hopkins, W. D. (2005). "Intentionality as measured in the persistence and elaboration of communication by chimpanzees (Pan troglodyes)," *Child Development*, 76 (1): 291–306.

Locke, J., 1689 [1997]. *An Essay Concerning Human Understanding*. London: Penguin Books.

Maple, T.L. and Perdue, B.M. (2013). *Zoo Animal Welfare*, New York: Springer.

Matsuzawa, T. (2010). "The chimpanzee mind: Bridging fieldwork and laboratory work," in E.V. Lonsdorf, S.R. Ross, and T. Matsuzawa (eds.) *The Mind of the Chimpanzee: Ecological and Experimental Perspectives*, Chicago, IL: The University of Chicago Press, pp. 1–19.

Mitani, J.C., Amsler, S.J., and Sobolewski, M.E. (2010). "Chimpanzee minds in nature," in E.V. Lonsdorf, S.R. Ross, and T. Matsuzawa (eds.) *The Mind of the Chimpanzee: Ecological and Experimental Perspectives*, Chicago, IL: The University of Chicago Press, pp. 181–191.

Osvath, M. (2009). "Spontaneous planning for future stone throwing by a male chimpanzee," *Current Biology*, 19 (5): R190–R191.

Osvath, M. and Karvonen, E. (2012). "Spontaneous innovation for future deception in a male chimpanzee," *PLoS ONE*, 7 (5): 1–8.

Perlman, J.E., Horner, V., Bloomsmith, M.A., Lambeth, S.P., and Schapiro, S.J. (2010). "Positive reinforcement training, social learning, and chimpanzee welfare," in E.V. Lonsdorf, S.R. Ross, and T. Matsuzawa (eds.) *The Mind of the Chimpanzee: Ecological and Experimental Perspectives*, Chicago, IL: The University of Chicago Press, pp. 320–331.

Perlman, J.E., Bloomsmith, M.A., Whittaker, M.A., McMillan, J.L., Minier, D.E., and McCowan, B. (2012). "Implementing positive reinforcement animal training programs at primate laboratories," *Applied Animal Behaviour Science*, 137 (3), 114–126.

Rakoczy, H., Clüver, A., Saucke, L., Stoffregen, N., Gräbener, A., Migura, J., and Call, J. (2014). "Apes are intuitive statisticians," *Cognition*, 131 (1), 60–68.

Rawls, J. (1971). *A Theory of Justice*, Cambridge, MA: The Belknap Press.

Ross, L. and Nisbett, R.E. (1991). *The Person and the Situation: Perspectives of Social Psychology*, Philadelphia, PA: Temple University Press.

Ross, S.R. (2010). "How cognitive studies help shape our obligation for the ethical care of chimpanzees," in E.V. Lonsdorf, S.R. Ross, and T. Matsuzawa (eds.) *The Mind of the Chimpanzee: Ecological and Experimental Perspectives*, Chicago, IL: The University of Chicago Press, 309–319.

Savage-Rumbaugh, S., Shanker, S.G., and Taylor, T.J. (1998). *Apes, Language, and the Human Mind*, New York: Oxford University Press.

Schechtman, M. (1996). *The Constitution of Selves*, Ithaca, NY: Cornell University Press.

Shumaker, R.W., Walkup, K.R., and Beck, B.B. (2011). *Animal Tool Behavior: The Use and Manufacture of Tools by Animals*, revised & updated edn., Baltimore, MD: The Johns Hopkins University Press.

Stanford, C. (2018). *The New Chimpanzee: A Twenty-First Century Portrait of Our Closest Kin*, Cambridge, MA: Harvard University Press.

Stumpf, R.M. (2011). "Chimpanzees and bonobos: Inter- and intraspecies diversity," in C.J. Campbell, A. Fuentes, K.C. MacKinnon, S.K. Bearder, and R.M. Stumpf (eds.) *Primates in Perspective*, 2nd edn., New York: Oxford University Press, pp. 340–356.

Tomasello, M. (2016). *A Natural History of Human Morality*, Cambridge, MA: Harvard University Press.

Tomasello, M. and Call, J. (1997). *Primate Cognition*, Oxford: Oxford University Press.

Varner, G.E. (2012). *Personhood, Ethics, and Animal Cognition: Situating Animals in Hare's Two-Level Utilitarianism*, Oxford: Oxford University Press.

Whiten, A. Goodall, J., McGrew, W.C., Nishida, T., Reynolds, V., Sugiyama, Y., Tutin, C.E.G., Wrangham, R.W., and Boesch, C. (1999). "Cultures in chimpanzees," *Nature*, 399 (6737): 682–685.

Wilson, E.O. (1985). "In the queendom of ants: A brief autobiography," in D.A. Dewsbury (ed.) *Leaders in the Study of Animal Behavior: Autobiographical Perspectives*, London: Associated University Presses, pp. 465–486.

Wittig, R.M., Crockford, C., Weltring, A., Langergraber, K.E., Deschner, T., and Zuberbühler, K. (2016). "Social support reduces stress hormone levels in wild chimpanzees across stressful events and everyday affiliations," *Nature Communications*, 7: 13361.

Wittgenstein, L. (1967). *Philosophical investigations*, 2nd ed. Oxford: Blackwell.

Wobber, V. and Hare, B. (2011). "Psychological health of orphan bonobos and chimpanzees in African sanctuaries," *PLoS ONE*, 6 (6): e17147.

Wolf, S. (2010). *Meaning in Life and Why It Matters*, Princeton, NJ: Princeton University Press.

Wolfensohn, S. and Lloyd, M. (2013). *Handbook of Laboratory Animal Management and Welfare*, 4th edn., New Delhi: Wiley-Blackwell.

Conclusions
Six

Under U.S. law the courts have a limited set of options as to how they regard chimpanzees such as Kiko and Tommy: they are either persons or things. The First, Third, and Fourth Departments of the Appellate Division of the New York Supreme Court have refused to recognize that Kiko and Tommy are persons, and so they remain things under law. Our contention has been that the reasons provided by the courts are either inadequate, or they actually support the personhood of Kiko and Tommy.

Perhaps, however, the paucity of the options is part of the problem. If a third option were available—that is, if there was some other way for the courts to approach the status of chimpanzees—then the decision of the courts might change. After briefly summarizing what we have argued in earlier chapters, we consider a third option, one some legal reformers and animal advocates have recently defended: the category of 'sentient being.' The introduction of this third category into the legal system might mean that chimpanzees such as Kiko and Tommy would not be limited to the two possibilities discussed above.

This third option has gained popularity in a number of European and Latin American countries, and a few American voices are calling for the U.S. to follow suit. However, even if it became a legal reality in the U.S. context, having this third option would not change our view that Kiko and Tommy are best regarded as persons under the law. But first, a brief recap of what we argued.

OUR ARGUMENT

In ruling against the NhRP and denying habeas corpus relief to Kiko and Tommy, the courts relied on an inconsistent and, at times, incoherent understanding of what it means to be a person. As we have shown, their rulings vacillate between competing interpretations of personhood, often within the same sentence, making it unclear what specific conception they were actually favoring. As philosophers, we set out to clearly differentiate the various conceptions deployed by the courts so as to investigate how each, when properly understood, bears on the question of Kiko's and Tommy's status as either persons or things.

To reiterate, our aim in this investigation was never to offer a positive account of personhood. Among the thirteen of us there is no consensus about the nature of personhood. Our aim as philosophers was simply to demonstrate that on any justifiable conception of personhood recognized by the courts, Kiko and Tommy are persons.

As the preceding chapters make clear, we were able to identify three different conceptions of personhood found in the courts' rulings, based on species membership, the social contract, or community membership. The courts fail to distinguish between these conceptions and often employ more than one. The fourth conception, capacities, is the one introduced by the NhRP in their petition.

We reject the species membership account of personhood as unjustifiable. This account confers moral and legal status upon the members of a morally irrelevant taxonomic classification and arbitrarily privileges *Homo sapiens*, while excluding members of other biological groups. Of the four accounts mobilized by the courts, this is by far the most deeply entrenched in the legal, social, and intellectual fabric of the U.S. (and elsewhere). That does not change the fact that the justification typically given in its defense is inconsistent with contemporary evolutionary biology.

We reject the suggestion that the social contract, which frames our legal and social institutions and expectations, creates persons. In the courts' view, only contractors can be persons. But this view is confused. Virtually all versions of social contract theory, classical and contemporary, understand that to be a contractor, one must be a person first, and insist that the real function of the social contract is to create and define citizenship, not personhood. Even if one grants the courts' assertion that Kiko and Tommy are excluded from the social contract (which is contested), this would be a reason to deny them citizenship but not personhood.

We address the courts' claim that personhood is a function of community membership and requires the right kind of connection to the human community. This view comes in two different but related versions: the personhood-by-proxy account (which links personhood to social recognition) and the social personhood account (which links it with being situated in webs of interpersonal relationships). On either version, we argued Kiko and Tommy are members of the relevant communities and so are persons.

We defend the capacities approach favored by the NhRP. On this account, individuals are persons if they possess one or more morally and legally relevant capacities, such as autonomy. Chimpanzees possess many of these capacities, including autonomy, and so qualify as persons.

At least two significant moral and legal implications follow from the argument developed in these four chapters:

a) Since under current U.S. law every entity must be categorized as either a person or a thing, Kiko and Tommy clearly should be categorized as persons.
b) Perhaps more decisively, there exists no conception of personhood that includes all humans who should be included, in the way they should be included, and excludes animals such as chimpanzees.

A THIRD CATEGORY: SENTIENT BEINGS

Some critics of chimpanzee personhood have argued that rather than fighting to secure legal personhood for chimpanzees, the NhRP should fight to expand the options available in the U.S. legal system to include a third category that better reflects our collective views about the legal status of nonhuman animals (Berdik 2013; Favre 2009). As they see it, the problem is not that we have historically failed to recognize the personhood of chimpanzees (and other animals), but that our legal system is not fine-grained enough to do justice to the place they occupy in our moral universe. The notion that objects in the world can be neatly pigeonholed into two broad categories—person and thing—is itself the problem because there are entities that are more than things but less than persons. Many nonhuman animals, and, presumably, no humans, occupy this interstitial territory.

The push for a third category has been a key component of various animal protection measures passed in Europe and the Americas. These reforms have used the term 'sentient beings' to denote entities whose legal status is taken to lie somewhere between thinghood and personhood. Consider the following recent examples:

- In February 2015, the French Parliament reformed the French Civil Code by adding Article 515-14, which defined animals as *des êtres vivants doués de sensibilité* or 'sentient beings'(Le Neindre et al. 2017: 13)
- In December 2015, the provincial government of Quebec passed Bill 54, an anti-cruelty statute that defined animals as 'sentient beings' and imposed heavy penalties on those found guilty of cruelty toward animals. It does not, however, change their status as property (The Canadian Press 2015: 1)
- In December 2015, the Congress of Colombia passed Bill 172, imposing some of the most stringent anti-cruelty protocols in Latin America. The law explicitly states that animals, as seres sintientes, or 'sentient beings,' "are not things". (Contreras 2016: 3)
- In February 2017, Mexico City ratified a new constitution that recognizes all animals as sentient beings "with inherent dignity," and stipulates that "all persons [in Mexico City] have an ethical duty and juridical obligation to respect the life and integrity of nonhuman animals," which "by their nature are subjects of moral consideration". (Ortega 2017: 1)

Similar reforms have passed in Germany, Austria, Brazil, Switzerland, and the Netherlands (Le Neindre et al. 2017: 13). We might speculate that these legal developments will yield improvements in animal welfare and protection and, perhaps, change social attitudes about the moral worth of animals, not by recognizing nonhuman animals as persons but by giving them a classification of their own: 'sentient beings.'

Not surprisingly, however, there has been disagreement about what the creation of a new legal category for animals means legally, socially, and politically. In France, for example, some have interpreted this third category as creating a whole new class of persons (so that one can differentiate between human and nonhuman persons), while others have interpreted it as creating a new class of things (so that one can differentiate between sentient and insentient property) (Le Neindre et al. 2017: 14).

To date, efforts to move from two to three legal options for classifying nonhuman animals have not been particularly successful in terms of actually changing animals' status as property. Indeed, these efforts sometimes further entrench that status, resulting in a confused and confusing legal status. The 2015 reform to the French Civil Code is a good illustration of this troubling trend. Though the reform recognizes that "animals are sentient living beings" it also states: "Aside from specific regulations that protect them, animals have to be legally treated as property" (Le Neindre et al. 2017: 13). Thus, the reform may be purely symbolic since it adds a new legal classification without invalidating the claim that animals are property under the law (Bolis 2014). The problem with this position is obvious. If you can treat an entity as a thing, it will remain a thing even if you call it by a different name.

An additional worry about this third category approach is that although it gives animals a new place in our legal systems, it may inadvertently encourage the continuation of an arbitrary division between humans and other animals by placing humans on a 'higher' moral level (see Chapters 1 and 2).

This worry notwithstanding, the third category approach may also benefit animals, and that matters ethically. Perhaps, for instance, it will succeed at giving animals fundamental rights (or certain protections akin to rights), such as the right to liberty or freedom from torture. If it does, then on matters concerning those rights (or protections), there may be no difference as far as the law is concerned between entities that we classify as 'sentient beings' and those we classify as persons. If it does not, then the ethical propriety of such a third category is in doubt.

Furthermore, we need to recognize the possibility that the introduction of this third category may not just affect the legal status of nonhuman animals. If we start regarding certain animals as sentient beings as a way of not recognizing them as persons, it is possible that such reassignments will not be limited to nonhuman animals. The creation of a category of beings who are sentient but not persons threatens to reinsert discriminatory hierarchies into philosophical and ethical discussions of personhood. It offers a possible legal mechanism for discriminatory attempts to re-classify certain humans from persons to 'sentient beings' under the law. In our analysis, we have taken pains to avoid these types of hierarchies.

FURTHER IMPLICATIONS

We end this book by encouraging the reader to think more broadly about where our argument might take us when it comes to the treatment and legal status of nonhuman animals in our societies.

First and foremost, humans are in the habit of making arbitrary and unjustified distinctions that profoundly and adversely affect the lives of other animals. We must think more critically about these distinctions. Our use of animals in biomedical research and our differential treatment of domestic animals illustrate the devastating effects of these unjustified distinctions.

There are numerous discriminatory legal regulations and policies that protect animals only relative to their location or use in greater society. For instance, animal welfare standards variously apply to such animals as birds, fish, horses, pigs, and rabbits, depending on whether they are companion animals or used in biomedical research or agriculture. As companion animals, they enjoy the kind of moral regard and legal protections afforded companion cats or dogs. As farm animals, or animals used in biomedical research, they do not. Our point here is not that common moral regard and extant legal protections for companion animals are ideal or adequate, but they are importantly different and offer more protection than is afforded to the very same animals when they are raised on farms for human consumption and, under law, are considered livestock.

These arbitrary distinctions conflict with our stance that a commitment to formal justice requires us to treat like cases alike. Of course, formal justice does not recognize the distinctness of individuals within or across species, but it does encourage us to pay attention to relevant *similarities* within and across species and to be vigilant about odious distinctions without difference that threaten the objectivity of our moral and legal norms. At the very least, a commitment to justice challenges us to reconsider the sphere of justice and to recognize that Kiko and Tommy belong within that sphere.

REFERENCES

Berdik, C. (2013). "Should chimpanzees have legal rights?" *The Boston Globe*, 14 July 2013. [Online] www.bostonglobe.com/ideas/2013/07/13/should-chimpanzees-have-legal-rights/Mv8iDDGYUFGNmWNLOWPRFM/story.html [Accessed 6 April 2018].

Bolis, A. (2014). "Les Animaux reconnus comme 'êtres sensibles,' un pas 'totalement symbolique," *Le Monde* [Online] www.lemonde.fr/planete/article/2014/04/16/les-animaux-reconnus-comme-des-etres-sensibles-un-pas-totalement-symbolique_4402541_3244.html [Accessed 6 April 2018].

Canadian Press (2015). "Quebec Defines Animals as 'Sentient Beings' in New Legislation," *CTV News*, [Online] www.ctvnews.ca/politics/quebec-defines-animals-as-sentient-beings-in-new-legislation-1.2687500 [Accessed 4 April 2018].

Contreras, C. (2016). "Sentient beings protected by law: Analysis of recent changes in Colombian animal welfare legislation," *Global Journal of Animal Law*, 2: 1–19.

Favre, D. (2009). "Living property: A new status for animals within the legal system," *Marquette Law Review*, 93: 1021.

Le Neindre, P., Bernard, E., Boissy, A., Boivin, X., Calandreau, L., Delon, N., Deputte, B., Desmoulin-Canselier, S., Dunier, M., Faivre, N., Giurfa, M., Guichet, J.L., Lasande, L., Larrère, R., Mormède, P., Prunet, P., Schaal, B., Servière, J., and Terlouw, C. (2017). "Animal consciousness," *EFSA Supporting Publications*, 14 (4).

Ortega, A. (2017). "In Mexico's capital, animals get constitutional consideration," *The Humane League Blog*, [Online] blog.thehumaneleague.org/mexico-city-sentient-beings [Accessed 1 April 2018].

Afterword
Steven M. Wise

Well-crafted philosophical arguments made by philosophers can change the world.

This book grew out of an amicus curiae brief that a group of seventeen North American philosophers filed in the New York Court of Appeals in February 2018. Their Philosophers' Brief urged the Court to grant a motion filed by the Nonhuman Rights Project, a civil rights organization dedicated to gaining fundamental legal rights for nonhuman animals that I founded in 1996. It sought review of a decision of New York's First Judicial Department, an intermediate appellate court, which had refused to issue an order to show cause under that state's habeas corpus statute on behalf of two captive chimpanzees named Tommy and Kiko. The sole substantive ground for the court's decision had been that Tommy and Kiko were not human beings. The court had also ruled against the NhRP on the procedural ground that it was not permitted to seek a second writ of habeas corpus after having been denied a first.

These were the NhRP's second motions for further review to the Court of Appeals on behalf of both Tommy and Kiko; we had asked the Court in 2015 to hear our appeal in Tommy's first case from a decision of another of New York's intermediate appellate courts that only a "person" could seek a writ of habeas corpus, that a "person" had to be able to bear duties as well as rights, and that chimpanzees could not bear duties. We had asked the Court to hear our appeal in Kiko's first case from a decision that said that no one, human or chimpanzee, was entitled to a writ of habeas corpus unless they demanded unconditional release from detention and the NhRP had asked that Kiko be sent to a sanctuary instead of being thrown onto the streets of New York.

The Philosophers' Brief did not address any legal errors the First Department might have committed. Instead it urged the Court of Appeals to hear the NhRP's further appeal because of errors the lower court had made in the area of their professional expertise: animal ethics, philosophy of biology, philosophy of animal cognition, and moral philosophy.

On May 8, 2018, the Court of Appeals refused to hear the NhRP's motion without giving any reason. This was unsurprising as that Court elects to hear fewer than 5% of the motions for further review that come before it. Notably, one of the judges, Eugene M. Fahey, issued a powerful and thoughtful opinion on the merits of the NhRP's claims. This was the first opinion by a judge of the highest court in any American jurisdiction that had ever addressed the issue of whether a nonhuman animal should be entitled to legal rights. Judge Fahey's opinion was clearly influenced by philosophical theories of personhood, some of which were contained within the Philosophers' Brief and some of which derived from philosophical sources outside the Brief that he consulted.

Judge Fahey attacked the idea that habeas corpus was reserved for those who can bear duties, noting that human infants and comatose human adults cannot bear duties,

> yet no one would suppose it was improper to seek a writ of habeas corpus on behalf of one's infant child

or a parent suffering from dementia. In short, being a 'moral agent' who can freely choose to act as morality requires is not a necessary condition of being a 'moral patient' who can be wronged and may have the right to redress wrongs (see generally Tom Regan, *The Case for Animal Rights* 151–156 (2nd ed. 2004)).

New York and other American judges have on occasion used the term "moral agent," and Judge Fahey got his definition right. However, no American court has ever used the term "moral patient" or cited any work of the influential animal rights philosopher, Tom Regan. In *The Case for Animal Rights*, Regan, whom I knew for decades and whose philosophical work heavily influenced my own work as a civil rights lawyer for nonhuman animals, defined "moral patient" as one who lacks the mental prerequisites that would allow her to be held morally accountable for what she does. "Moral patients, in a word, cannot do what is right, nor can they do what is wrong." Regan noted that some moral patients are conscious and sentient, while others, including chimpanzees, possess additional cognitively complex abilities.

Judge Fahey observed that

> (t)he Appellate Division's conclusion that a chimpanzee cannot be considered a 'person' and is not entitled to habeas relief is in fact based on nothing more than the premise that a chimpanzee is not a member of the human species ... I agree with the principle that all human beings possess intrinsic dignity and value ... but in elevating our species, we should not lower the status of other highly intelligent species.

The question for the court, he wrote, should be whether a chimpanzee has the right to liberty protected by habeas corpus and its answer "will depend upon our assessment of the intrinsic nature of chimpanzees as a species."

The judge noted both the unrebutted extensive affidavits filed by eminent primatologists, which demonstrated chimpanzees' cognitively complex minds, and that "the amici philosophers with expertise in animal ethics and related areas draw our attention to recent evidence that chimpanzees possess autonomy by self-initiating intentional, adequately informed actions, free of controlling influences." He therefore concluded that

> (t)o treat a chimpanzee as if he or she had no right to liberty protected by habeas corpus is to regard the chimpanzee as entirely lacking independent worth, as a mere resource for human use, a thing the value of which consists exclusively in its usefulness to others. Instead we should consider whether a chimpanzee is an individual with inherent value who has the right to be treated with respect (see generally Regan, *The Case for Animal Rights* 248–250).

Moreover, the refusal of courts to consider these issues "amounts to a refusal to confront a manifest injustice."

After confessing that he has "struggled" with whether he should have voted to deny the NhRP's appeal the first time the Tommy and Kiko cases came before the Court of Appeals in 2015 and that he has continued to question the correctness of his decision, Judge Fahey concluded with this assertion:

> The issue whether a nonhuman animal has a fundamental right to liberty protected by the writ of habeas

corpus is profound and far-reaching. It speaks to our relationship with all the life around us. Ultimately, we will not be able to ignore it. While it may be arguable that a chimpanzee is not a 'person,' there is no doubt that it is not merely a thing.

The relationship between legal justice and moral ethics is both complex and important. Judges are specialists in issues of legal justice. But, as Judge Fahey's opinion demonstrates, at least with respect to such an issue as whether nonhuman animals deserve legal rights, the considered opinions of moral philosophers can assist them in coming to the morally right and legally correct decisions. And those decisions – for example that of Lord Mansfield to end human slavery in England in the famous 1772 case of *Somerset v. Steuart* – do indeed change the world.

Author index

Bard, K. 86
Beauchamp, T. L. 83–4
Blackstone, W. 34n3

Carruthers, P. 56–8
Christman, J. 80
Crary, A. 8
Cupp, R. 54–6, 58

Darwin, C. 20–2, 24, 33
DeGrazia, David 32
de Waal, Frans B.M. 85, 87, 90–1

Fahey, E. M. 112–15

Gallup, G. 87–8
Goodall, J. 84, 86

Harris, Angela P. 5–6
Hercules 1–2
Hobbes, T. 41, 43–7

Kant, I. 41–2, 80
Kim, C. 19–20
Ko, A. 5–8
Ko, S. 5–8
Korsgaard, C. 81–2

Linnaeus, C. 17–18, 21, 29
Locke, J. 43–5, 47

Mills, C. 18

Narveson, J. 62
Nisbett, R. 82
Nishida, T. 84
Nussbaum, M. 52–3

Rawls, J. 41–2, 49–52
Regan, Tom 113–14
Ross, L. 82
Rousseau, J.J. 43–5, 47–8
Rowlands, M. 42, 49, 51, 53

Salmond, J. W. 49
Sarah xiv–vi, xviii–ix
Schiebinger, L. 18

Taylor, S. 7
Tomasello, M. 87, 89
Tommy xiii, xvii–ix, 1–5, 9–10, 13, 34, 41–2, 44–5, 49, 54, 59, 61, 66–71, 73–4, 94–5, 101–4, 108, 111, 114

Wilson, E. O. 83
Wise, S. 2, 7, 111–15
Wobber, V. 83–4
Wynter, S. 6

Subject index

NOTE: page numbers followed by 'n' refer to endnotes.

ableism/ableist 5, 7, 8, 16
adequate sanctuary environment 89, 92–3
Aibo robot dogs (community membership) 71–2
amicus brief, amicus curiae xviii, 2–4, 10, 111
animality and race (Ko) 6–7
animality and disability 7
Appellate Court (New York Supreme Court) 111
autonomy 29, 45, 54, 62, 66, 73, 74n3, 77–8, 95n11, 103, 114; adequate sanctuaries 89, 92–3; choice and control 89–93; and ethics 93–4; evidence 83–5; personhood and 80–3; self-governance 80–3, 91; welfare 89–92; *see also* capacity for autonomy

biological taxonomy 16–24
biomedical laboratories 93
Black's Law Dictionary 49

capacity for autonomy 77–8; evidence *see* evidence; of personhood 78–9; *see also* autonomy

captivity xiii, 1, 10n2, 59, 70, 94
Chimp Haven 92–4
choice and control 89–93
chimpanzees xiii–xix, 1, 2, 5, 8, 10, 14, 20, 29, 31, 36n11, 48, 55, 67–70, 77–8, 83–94, 96n16, 101, 103–4, 111–15
citizens 43–4
cluster version 77, 79
cognition xv, xixn5; cognitive 1, 7, 30, 66, 74n1, 84, 89, 91
Commentaries on the Laws of England (Blackstone) 34n3
Common Law 1, 4, 18
communication 29, 86
community membership 61–2; implications 66–70; limits 70–6; personhood-by-proxy 62–3; social personhood 63–6
companion animals 108
contractarianism 51, 58; contractarianism (Hobbes) 41, 43–7, 52
"Contractarianism and Animal Rights" (Rowlands) 51
contractualism 51–2, 58; contractualism (Kant) 41–2, 52, 80

Declaration of Independence 46
disability 5, 7, 16, 24, 63, 65, 74n1
duties (legal and moral) 42, 46, 48–9, 59, 61, 65, 112

emotions 73, 78, 84–6
essentialism 13, 15–16, 25–6
essentialist version 77, 79
ethics 80, 89, 93–4, 115
evidence 84; autonomy 85; emotions 85–6; language 86; rationality 87; self–awareness 87–8; sentience 87; sociality 88–9
evolution 20–2, 24–9, 31–2, 35n7, 35–6n10; evolutionary theory 25–7, 103

First Department of the Appellate Division of the New York Supreme Court 9, 10n1, 48, 61, 66–7, 112
Fourth Department of the Appellate Division of the New York Supreme Court 10n2, 93
free-living chimpanzee community 68–70, 89–91
Frontiers of Justice (Nussbaum) 52–3

genetic difference 31
Great Chain of Being 17, 19–20, 23, 34n4

habeas corpus 1–2, 4–5, 9–10, 13, 55, 59, 66, 69, 93, 102, 111–12, 114
harm 7, 16, 52, 65, 67, 72, 89, 92–3, 95–6n13
Hominidae 14, 34n1
Homo floresiensis 32
Homo neanderthalensis 32

Homo sapiens 6, 13–16, 23, 27–34, 51, 55, 61, 73, 95, 103
human: characteristics 31–3; and chimpanzees 29; homo sapiens and 27–33; nature 30–1; rights 56–8; and species membership 28–30
humanity 19, 30, 54–5
hybridization 23, 32, 35n10

jurisprudence (Salmond) 49
justice 4, 34, 50–3, 55, 59, 108, 115

Kiko xiii, xvii–ix, 1–5, 9–10, 13, 34, 41–2, 44–5, 49, 54, 59, 61, 66–71, 73–4, 93–5, 101–4, 108, 111–12, 114

language xv, 30, 78, 82, 86
Leo 1–2

marshmallow experiment 85
meaning-making 79
minds 114; the mental xv, 73, 78, 113
mirror-self recognition 87–8
morality 30, 79, 113

narrative self-constitution 78
natural kinds 15–16
natural rights 45–8
Neanderthals (*Homo neanderthalensis*) 32
New York Court of Appeals ii, 2, 111, 114
New York Supreme Court 9, 10n2, 101
Nonhuman Rights Project (NhRP) 1, 9–10, 14, 34, 44, 49, 54, 61, 67, 77, 94, 102–4, 111–12

oppression 5–8, 24, 57

personhood: and autonomy 80–3; conceptions 102–4; distinctions conflict 107–8; legal persons/personhood xviii, 2–3, 7, 13–14, 16, 32–3, 34n3, 48–9, 54–5, 57, 61–2, 64–5, 74n1, 79, 94, 102–4, 106–7; sentient beings 104–7

personhood-by-proxy 57, 62–3, 66–7, 70–2

philosophers xviii, 2–4, 15, 23, 95, 102, 111–12, 114–15

philosophy 3, 41, 50, 62–4, 72, 78–9

population thinking 23

race xvii, 5–7, 16, 18–19, 21–2, 24, 34n4, 50–2, 65, 74n1

racism/racist 5–8, 16, 19, 22

rationality 29–30, 44–45, 50, 52, 54, 56–7, 78, 87

relational personhood 62–3

responsibilities (legal and moral) 42, 44–6, 48–9, 61–2, 64–5, 72, 81–2

rights 1, 4–9, 13, 16, 18–20, 24, 32–4, 41–9, 54–9, 74n1, 107, 111–13, 115

sanctuaries xiii–xiv, xix, 92–3

self-awareness 28, 78, 84, 87–8, 95n11

self-consciousness 81

self-standing persons 62–7, 70–1

sentience 72, 78, 84, 87

sentient beings 104–7

sex 16–19, 21, 24

sexism/sexist 16, 18

social contract 41–3; arguments 54–8, 103; contractor 44–5; legal person 48–9; natural rights 45–8; nonhuman personhood 49–53; personhood and citizenship 43–4

sociality 78, 88–9

social justice struggles 5, 7–8

social personhood 63–70, 72

Somerset v. Stewart 7, 115

species membership 13–14; biological taxonomy, history of 16–24; essentialism 15–16; evolutionary theory 25–7; humans and *see* humans; natural kinds 15–16

Species Problem 26–30

speciesism/speciesist 55

State of Nature 45, 47–8, 52

A Theory of Justice (Rawls) 50

thinghood 2–3, 69, 101–2, 104–6

Third Department of the Appellate Division of the New York Supreme Court 9, 42, 43, 46, 48

Tree of Life 14, 20, 23, 26, 33, 34n4

Unlocking the Cage 2

Washoe 86

zoos xiii, 71, 91, 93

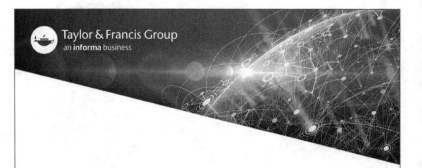